A LIVING FACTORY

Every one of the cells in our body is a chemical factory; inside each pinpoint of life thousands of different chemical reactions are occurring constantly. Through these reactions, food is digested and tissue is built.

This amazing activity is described fully as the author identifies the substances essential to human life: proteins, the elemental ingredient without which life in any form cannot exist; enzymes, the catalytic proteins which control body chemistry; hormones, the compounds that dictate the function of the enzymes; vitamins, the important substances which must be obtained from sources outside the body.

The reader learns why he must eat certain foods, what happens chemically when he is alert or fatigued. Mysterious chemical contradictions are revealed—poisons kill in a matter of seconds by inhibiting the vital function of the enzymes . . . wonder drugs, on the other hand, often act in the same chemical manner but with life-saving results.

Here is the full picture of the chemistry of life, the constant breakdown and rebuilding of the moving, living organism.

The Chemicals of Life

by ISAAC ASIMOV

A MENTOR BOOK
NEW AMERICAN LIBRARY
TIMES MIRROR
NEW YORK AND SCARBOROUGH, ONTARIO

*To my
friends
and
colleagues:*

BURNHAM S. WALKER
and
WILLIAM C. BOYD

Contents

Illustrations

The
Chemicals
of
Life

1

The All-Important Protein

ALL LIFE IS PROTEIN

IF WE WERE to study the human body very closely, we would find it to contain a number of different chemical substances. In the first place, we would find water; a great deal of it. About three-fifths of the body is water. Quite ordinary water, too; the kind you can find in any reservoir.

In the bones and teeth are minerals quite similar in nature to certain ordinary types of rock. In the blood there is a small quantity of sugar; not exactly the kind we buy in the store, but closely related to it. In the liver, there is a kind of starch which isn't very different from potato starch or corn starch. And in many parts of the body, especially under the skin and about the kidneys, there is fat.

It would seem impossible that life could be built up out of these commonplace substances. Actually, it *is* impossible, since one all-important ingredient of the human body has been omitted from the list. In every part of the body, there is a type of substance called PROTEIN. Some parts of the body contain more than others. Blood and muscles are one-fifth protein; brain is one-twelfth protein; tooth enamel is less than one-hundredth protein. The point is that no living portion of the body is completely without protein.

This holds true not only for human beings, but for all plants and animals. There is no living organism at all, of any kind, that does not contain protein. Even if we consider very small and simple forms of life, such as bacteria, we find that each individual bacterium contains proteins. If we probe deeper still, to the very simplest objects we can call alive, the viruses, we still find protein.

Viruses are so small that bacteria are giants in comparison. Some are so small that it would take a million of them in a row to make a line one inch long. They are so small that they have given up almost all the functions of life, except the ability to multiply. They do this inside the living cells of other creatures, and in the process frequently cause disease. Measles, infantile paralysis, and the common cold are examples of diseases caused by virus multiplication in the cells of human beings.

The smallest viruses seem to be composed only of chemicals that are absolutely necessary to life. They have no room for non-essentials. It is important to note, therefore, that such viruses are composed of nothing more than a particularly complicated kind of protein called nucleoprotein.

Nucleoprotein consists of a type of substance called *nucleic acid* in association with protein. It is nucleic acid that is now supposed to control the behavior and properties of cells. It does this, however, by means of the particular proteins it causes to be formed.

When proteins were first isolated from living tissue about a hundred and twenty years ago, scientists realized almost at once that it was something quite special. Even its name shows that. A Dutch biochemist, named Mulder, who first used the word "protein" back in 1838, got it from a Greek word meaning "holding first place."

Protein certainly does that, as far as life is concerned.

In this book, we are going to explain the reasons why proteins are important, and then we will discuss the vital role played by a group of proteins known as ENZYMES.

THE INGREDIENTS OF PROTEINS

Well, what are proteins made of?

These days, almost everybody knows that everything is made up of ATOMS. Atoms are so small that the tiniest virus can hold millions of them. Proteins are made up of atoms, and so are the rocks, the stars, and, in fact, everything.

Not all atoms are the same. There are various kinds of atoms. These behave differently in many ways so that the chemist has learned to tell them apart by methods which have been continuously improved for the last two hundred years. The chemist has given a different name to each variety of atom, and these different varieties are called ELEMENTS. Some of these elements, like gold, silver, iron, and copper, are familiar to all of us. Others are so unusual that very few people other than professional chemists have even heard of them.

One might think, considering the immensity of the universe and the huge variety of things in it, that there must be a vast number of different elements. This is not so! Altogether, 103 different elements are known today and of these most occur very rarely. Actually, a dozen different elements and no more make up over 99 per cent of all we see, from our own noses to the furthest star.

Atoms rarely occur by themselves. They seem to prefer to exist as groups of varying size. Such a group of atoms is called a MOLECULE. The substances we see all about us have the properties of the molecules of which they are composed, rather than of the individual atoms. While there are only 103 kinds of atoms there are thousands upon thousands of different kinds of molecules, since even a few kinds of atoms can group themselves in many different ways. When a molecule is made up of a group of two or more *different* kinds of atoms, the substance containing that molecule is known as a COMPOUND.

Compounds can be worlds different from the elements which make them up. For instance, the water molecule is composed of three atoms, two of which are hydrogen and one of which is oxygen. Now hydrogen and oxygen are both gases. Hydrogen is a very light gas; the lightest gas known, in fact. It was once used to fill airships in order that they might float in air. The trouble with that was that hydrogen can catch fire very easily and several horrible accidents resulted from its use. Oxygen forms part of the air about us. In fact, one-fifth of the air is oxygen and it is the oxygen portion which is necessary to life. Land animals breathe in order to suck oxygen into their lungs and thus make it available for use by the body.

Now it is obvious that water, although it is made up of hydrogen and oxygen, is completely different from those gases. In the first place, water is not a gas under ordinary conditions, but a liquid. It will not catch fire, like hydrogen, and it cannot be breathed, like oxygen. (Even fish don't breathe water itself. They simply suck water past their gills and take out the small quantities of dissolved oxygen that the ocean has absorbed from the air.)

The water molecule is very simple as molecules go. An example of a more complicated one is the molecule of ordinary table sugar which contains no less than 45 atoms of three different kinds. There is the hydrogen and oxygen which we have just considered, and a third element, carbon, in addition.

Carbon is just about the most versatile element that exists. It is familiar to all of us in several forms. Hard coal is almost pure carbon. So is the "lead" (more properly called graphite) of soft lead pencils. So is the soot which settles on the inside of chimneys. In all three cases, carbon is a black, dull material. How surprising then to find out that still another form of carbon is the brilliant and sparkling diamond. There certainly seems to be a great difference between diamond and coal and no young lady would be pleased to have her engagement ring set with a small piece of graphite. Nevertheless, all these are different forms of carbon. The difference between them rests entirely in the way in which the carbon atoms are arranged. The carbon arrangement is very tight in diamond and looser in coal and graphite.

As though this were not enough, carbon is versatile in other ways as well. Whereas the atoms of most elements prefer to combine with other atoms only in a limited way to form small molecules, carbon is quite different. Carbon atoms will combine among themselves and with atoms of other elements to an almost unlimited extent. Molecules that contain carbon can therefore exist in all shapes, sizes and varieties. There are far more different carbon-containing compounds known than the number of all other compounds of the remaining 102 elements.

Carbon can combine with hydrogen in various arrangements to form such substances as illuminating gas, gasoline, kerosene, dry-cleaning fluids, cigarette lighter fuel, asphalt, and vaseline. Carbon can combine with hydrogen and oxygen to form sugar, starch, wood, and fat. Finally carbon can combine with hydrogen, oxygen, nitrogen and sulfur to form protein.

Two new elements have just been mentioned and a word about each is in order. Nitrogen is a gas, which, like oxygen, is found in the air. In fact, it forms the major part of the air, making up nearly four-fifths of it. Unlike oxygen, it is of no use to the body as a gas. It enters the lungs when we inhale and leaves again when we exhale without being in any way affected by its experience. As part of compounds, however, nitrogen is very useful. It is an essential part of fertilizers, and of explosives such as TNT and nitroglycerine, as well as of proteins.

Sulfur is a yellow solid which can burn in air with a blue flame and a choking odor. In combination with hydrogen, it forms a gas which has an unpleasant odor of another type, that usually reminds us of rotten eggs.

Now we can answer the question with which this section started. Protein molecules always contain atoms of carbon, hydrogen, oxygen, and nitrogen. They usually contain sulfur as well. Sometimes they also contain other kinds of atoms, but we will get to that later on in the book.

THE SIZE OF PROTEINS

What makes proteins so unusual? Well, for one thing, the protein molecule is very large. To show what we mean by that, let's consider the weight of different kinds of atoms and molecules.

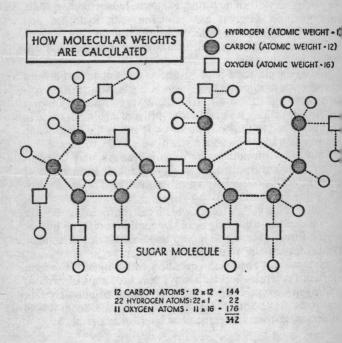

HOW MOLECULAR WEIGHTS ARE CALCULATED

○ HYDROGEN (ATOMIC WEIGHT · 1)
◉ CARBON (ATOMIC WEIGHT · 12)
☐ OXYGEN (ATOMIC WEIGHT · 16)

SUGAR MOLECULE

```
12 CARBON ATOMS · 12 x 12 = 144
22 HYDROGEN ATOMS: 22 x 1  =  22
11 OXYGEN ATOMS · 11 x 16  = 176
                            ―――
                            342
```

Naturally, all atoms are exceedingly light. It takes billions upon billions of them to make up the weight of even the tiniest particle of dust. It is one of the miracles of science that man has been able to weigh atoms despite their minuteness.

Now it turns out that the hydrogen atom is the lightest

one that can exist and it is customary to call its weight 1 for convenience. Or, to put it just a little more scientifically, 1 is called the ATOMIC WEIGHT of hydrogen. The carbon atom is 12 times as heavy as the hydrogen atom and carbon's atomic weight is therefore 12. In the same way, we can say that the atomic weight of nitrogen is 14, of oxygen is 16, and of sulphur is 32.

In order to find out how much a molecule weighs, it is only necessary to add up the atomic weights of the various atoms it contains. For instance, the hydrogen molecule consists of two hydrogen atoms, each with an atomic weight of 1. The MOLECULAR WEIGHT of hydrogen is therefore 2. Similarly, the nitrogen molecule is made up of two nitrogen atoms which weigh 14 each. The oxygen molecule is made up of two oxygen atoms which weigh 16 each. The molecular weight of nitrogen is therefore 28 and that of oxygen is 32.

The same rule holds where the atoms in a molecule are of different types. The water molecule, with one oxygen and two hydrogen atoms, has a molecular weight of $16 + 1 + 1$, or 18.

As we said in the previous section, the water molecule is a rather small one. A molecule of table sugar, by contrast, has 12 carbon atoms, 22 hydrogen atoms and 11 oxygen atoms. The 12 carbons weigh 144 altogether, the 22 hydrogens weigh 22, and the 11 oxygens weigh 176. Add them all together and the molecular weight of sugar turns out to be 342. This is a more sizable figure than that for water, but it is by no means tops. A molecule of a typical fat contains as many as 170 atoms and has a molecular weight of nearly 900.

Now we are ready to consider the protein molecule. How does it compare with fat and sugar in this respect? Of course, there are innumerable different kinds of protein molecules, but we can pick a protein that occurs in milk and has been studied quite a bit. *In its molecule are no less than* 5,941 *atoms.* Of these, 1,864 are carbon, 3,012 are hydrogen, 576 are oxygen, 468 are nitrogen and 21 are sulphur. The molecular weight, as you can see for yourself, is quite large. It comes to 42,020. The molecule of this protein is thus 45 times as large as a molecule of fat and 120 times as large as a molecule of sugar.

SIZES OF MOLECULES

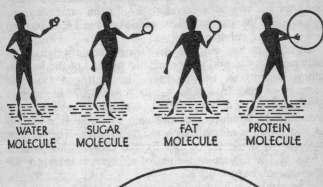

WATER MOLECULE   SUGAR MOLECULE   FAT MOLECULE   PROTEIN MOLECULE

VIRUS MOLECULE

But is this protein a fair example? Actually, it is not, because it is a rather *small* protein. The average protein has a molecular weight of 60,000. Many go much higher. Some of the proteins in clam-blood, for instance, have a molecular weight of 4,000,000. And some of the viruses consist of protein molecules with molecular weights in the tens of millions and even the hundreds of millions.

Now size in itself can be very useful. The body can do things with a protein molecule that it could not do with smaller molecules. It is as though you were given the choice of having a birthday party in the large ballroom of an expensive hotel or in a little one-room tenement flat. Obviously the ballroom would have many more possibilities (provided money were no object).

THE COMPLEXITY OF PROTEINS

But is size alone enough? One could imagine a large ballroom with no furniture and no ventilation. It might then be preferable to give the birthday party in the small flat after all.

Actually, there are molecules that are just as large as proteins but that are nevertheless much more limited in their usefulness than proteins. For instance, the chief compound of ordinary wood is CELLULOSE. Its molecule is very large but its only use to the plant is as a stiffening substance in the "walls" around the living plant cells. Again, the starch-like substance called GLYCOGEN, that occurs in animal livers, has a large molecule and yet is used only as a body fuel. Proteins, on the other hand, have millions and billions of different functions in the body.

Why is this so? Well, the key to the mystery can be found if cellulose or glycogen is treated with certain acids. These acids cause the cellulose or glycogen molecule to break up into smaller pieces. The smaller pieces turn out to be the same in both cases. They are molecules of GLU-COSE, a kind of sugar which is found in blood and which is somewhat simpler than ordinary table sugar.

The cellulose molecule, in other words, seems to resemble a necklace made up of thousands of individual glucose molecules strung together like so many beads. The glycogen molecule is made up of these same glucose molecules strung together in a somewhat different pattern.

Apparently, the fact that cellulose and glycogen are made up of only one type of smaller molecule limits their versatility. This also holds true for other such giant molecules *(with the exception of proteins)* which almost always consist of only one (or sometimes two) sub-units. It is as though you were given the job of making up a language but were only allowed to use a single letter. You could have words like *aa* and *aaaa,* and *aaaaaaaaaaaaaa.* In fact, you could have any number of words, depending on how many *a*'s you wished to string together, but it wouldn't be a satisfactory language. Things would be a little better if you were allowed to use two letters; still better if allowed to use three; and very much better if allowed to use twenty.

The last is exactly the case in proteins. When proteins are exposed to acid, their molecules also break apart into a number of smaller molecules. These smaller molecules are known as AMINO-ACIDS, and they are *not all the same.* There are about twenty different amino-acids, varying in size from a molecular weight of 90 to one of about 250. They can be strung together to form proteins in every which way. And each time they are strung together in a slightly different order, they make a slightly different protein.

How many different combinations are there possible in a protein molecule? Well, an average protein molecule would contain about 500 amino-acids, altogether, but we can start with a much smaller number. Suppose we start with only two different amino-acids and call them *a* and *b*. They can be arranged in two different ways: *ab* and *ba*. If we had three different amino-acids, *a, b,* and *c,* we could make six combinations: *abc, acb, bac, bca, cab,* and *cba*. With four different amino-acids, we could make 24 different combinations. They are easy to figure out and the reader may wish to amuse himself by listing them.

However, the number of possible arrangements shoots up very sharply as the number of amino-acids is increased. By the time you get to ten different amino-acids, there are more than 3,500,000 possibilities and with twenty amino-acids, almost 2,500,000,000,000,000,000 arrangements. (This seems unbelievable, but it is so. If the reader is doubtful, let him try listing the different arrangments for only 6 amino-acids. He will probably give up long before he has run out of arrangements.)

In the case of our average protein with 500 amino-acids, even though the 500 are not all different, the number of possible arrangements is so large that it can only be ex-

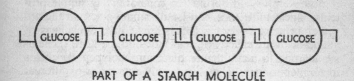

HOW LARGE MOLECULES ARE BUILT UP

PART OF A STARCH MOLECULE

PART OF A PROTEIN MOLECULE

pressed by a 1 followed by 600 zeroes. This is a far, far greater number than the number of all the atoms in the universe. You may understand why this should be if you will imagine taking the 26 letters of the alphabet and counting the number of words you can make out of them. Not only the real words now, but words with any number of letters up to 500, and especially including all the unpronounceable ones.

Remember that each one of these amino-acid arrangements is a slightly different protein. It is no wonder, then, that the body can design different proteins to accomplish different tasks without any danger of ever running out of new varieties. No wonder, too, that out of a type of molecule such as this, life can be built.

SOME FAMILIAR PROTEINS

You may be wondering if you have ever seen a protein. If you are, rest assured that you have. The hair on your head is an example of an almost pure protein. So is silk. The protein of hair is called KERATIN by chemists, and the protein of silk is called FIBROIN.

Both keratin and fibroin are comparatively simple proteins. Their molecules consist of amino-acids strung together in more or less a long straight line. Such lines of amino-acids are called POLYPEPTIDES. In the 1940's, chemists learned to manufacture quite long polypeptide chains in the laboratory. They used only one or two different amino-acids in doing so, however.

Then, in the 1950's, chemists learned how to put together amino-acids of many different varieties, one by one, just in a particular order. By 1960, a protein built up of 23 amino-acids was manufactured in the laboratory. It was found to behave just like a similar small molecule formed by the body. However, 23 amino-acids is a long way from the hundreds and thousands of amino-acids found in the larger proteins made by the body.

Still, fibroin isn't much more complicated than these laboratory creations. Its molecule consists of over 350 amino-acids of 14 different kinds, but actually 11 of the varieties occur to the extent of only a few each. Eighty-five per cent of the molecule is made up of only three different amino-acids, and those three happen to be the simplest of all. It is for this reason that silk doesn't play a vital role in life. It is just used by the silkworm to make a soft cocoon for itself.

Proteins such as fibroin and keratin are called FIBROUS PROTEINS. In general, fibrous proteins are strong, sturdy and tough and are used where the body needs support or protection. Keratin, for instance, is the chief protein not only of hair, but of skin, nails, hooves, scales, horns and feathers. Another important fibrous protein is COLLAGEN, which occurs in cartilage, ligaments, and tendons.

The really important proteins, however, are the GLOB-ULAR PROTEINS. In these, the polypeptide chains are not merely straight lines, but exist in complicated loops and twists which are never quite the same in any two different proteins. It is the globular proteins that do the main work of living and among them are the powerful and mysterious substances known as enzymes, which we will begin discussing in the next chapter.

SUMMARY

The chemicals of the body which are most important to life are called proteins. Proteins are made up out of five different substances: carbon, hydrogen, oxygen, nitrogen and sulphur. The protein molecule is very large, being hundreds of thousands of times greater than the molecules of such substances as water, for instance. The protein molecule is made up of twenty different varieties of smaller molecules called amino-acids. These amino-acids are strung together in a line, like pearls on a necklace, to form proteins. Since they can be strung together in many different ways, there are many different kinds of proteins.

2

Enzymes: Proteins that Make Haste

THE MAGICAL CATALYST

ONE OF THE FAMILIAR BOTTLES in the medicine cabinet
is the one labeled hydrogen peroxide. Actually, the peroxide we buy at the drugstore is not the pure chemical.
Pure hydrogen peroxide would be far too strong to use
on cuts and scrapes. The liquid we do use is only 3 per
cent hydrogen peroxide. The remaining 97 per cent is
just ordinary water. When hydrogen peroxide is mentioned
here, however, the pure chemical is meant.

Hydrogen peroxide is a close relative of water. The
molecule of water consists of two hydrogen atoms and
one oxygen atom, while the molecule of hydrogen peroxide contains two hydrogen atoms and *two* oxygen atoms.
The second oxygen atom, however, is not in a comfortable
position. The combination of two hydrogens and one oxygen is a very congenial one. A second oxygen is an in-
truder. Consequently, the hydrogen peroxide molecule
does what it can to let the second oxygen escape so that
it may become a good, reliable water molecule. This is
what chemists mean when they say that hydrogen peroxide
is "unstable."

Now there is no easy way of stopping this hydrogen peroxide breakdown, or "decomposition" as it is more properly called. As the peroxide bottle stands, the chemical gets weaker and weaker until finally it is useless. The best that can be done is to slow up the decomposition. Thus, strong light encourages the hydrogen peroxide molecules to get rid of their extra oxygens faster. The chemical is therefore always kept in a brown bottle, the glass of which lets through very little light. Again, hydrogen peroxide breaks down faster and faster as the temperature goes up. For that reason, if you will look at the hydrogen peroxide bottle, you will notice that the label says, "Store in a cool place."

The oxygen which is set free when hydrogen peroxide breaks down is a strong chemical. It attacks and kills many germs, so that hydrogen peroxide can be used as an antiseptic. It will also combine with various colored substances and destroy the color. This process is known as "bleaching." It will work on hair, for instance (DON'T TRY IT YOURSELF), and perhaps you have heard the expression "peroxide blonde."

In addition to heat and light, there are other ways of hastening hydrogen peroxide breakdown. Some iron filings added to hydrogen peroxide will cause oxygen to come out of the molecules so fast you can actually see bubbling. (DON'T TRY THIS AT HOME.)

Why should this be? The iron doesn't appear to be doing anything. It isn't changed in any way as a result of its experience. After the hydrogen peroxide is all broken down, the liquid can be poured off, and the iron filings can be used to decompose more of the chemical. It can be used for as many bottles as you wish.

This is not the only example of such an action of one chemical upon another. There are thousands of thousands of cases where a small pinch of some substance will hasten the change of great quantities of one molecule into equally great quantities of another.

Such substances that seem to influence other molecules by their mere presence without themselves being changed are known as CATALYSTS. They are said to CATALYZE various chemical reactions.

Industry makes great use of catalysts of various

sorts. One of the best known catalysts is powdered platinum. This, as you can well believe, is quite expensive. Consequently, powdered nickel (a metal which occurs in the American five-cent coin and gives it its name), which is almost as good, is usually substituted.

Here is an example of the use of nickel. Certain plant oils, such as cottonseed oil, have unpleasant tastes and cannot be used in cooking. If treated with hydrogen gas, these oils will harden into solid fats that are satisfactory for kitchen use. They are also cheaper than animal fats, such as lard or butter. The process of hardening is, unfortunately, very slow, but a small quantity of powdered nickel speeds it up tremendously. It is to a catalyst, then, that we owe such shortenings as Crisco and Spry.

Water itself is a catalyst for many processes. It was stated in the previous chapter that hydrogen burnt easily in air. This only happens, however, if there is a little water present. If the hydrogen and the air are both *absolutely* dry, hydrogen will not burn. The introduction of a small drop of water, however, could cause an instant explosion, if the temperature were high enough.

In many respects, the catalyst seems to be almost magical in its properties. How amazing it is, then, that our body is full of thousands of different catalysts which it manufactures for itself.

You may have observed the workings of one of the body's catalysts for yourself. At least you have if you have ever put hydrogen peroxide on a cut. As soon as that chemical touches the exposed tissue it froths. The hydrogen peroxide molecules break down so rapidly that the liquid fairly turns white with oxygen bubbles.

The body catalyst that does this has an appropriate name. It is called CATALASE.

Catalase is just one of a large group of body catalysts. Each member of the group speeds up some particular chemical change just as catalase hastens the decomposition of hydrogen peroxide. They make haste, in other words. The group as a whole are known as enzymes. It is these enzymes which form the main subject of this book.

WHY A LITTLE BIT
OF ENZYME GOES A LONG WAY

How does a catalyst work? Obviously, it isn't magic, no matter how magical it may seem.

There are two general ways in which a catalyst can do its job. In the first place, it may start a CHAIN REACTION. What we mean by that can be explained by considering a piece of burning paper. Paper, ordinarily, will not start to burn by itself. However, if it is raised to a high temperature, say by the flame of a match, it will burn briskly. Perhaps only one small corner of a large sheet of paper is set afire by the match. As that corner burns, however, the heat it develops is sufficient to set the neighboring regions of the paper afire. The new heat will set fire to additional portions of the paper, and the process will proceed on and on, faster and faster, until all the paper is burnt. No second match is necessary. By the same chain reaction process, in which each step is a link to the next step, a single smoldering cigarette stub can burn down a forest.

The action of water in promoting the burning of hydrogen is an example of a catalyst that works by starting a chain reaction. The details involve complicated chemistry, but, fortunately, they needn't bother us now.

Another way in which a catalyst may work is to supply a surface on which a certain reaction can take place very easily. Thus, a molecule of hydrogen peroxide may manage to stick to the surface of a piece of iron. It then finds that it can kick the extra oxygen loose more easily than it could otherwise.

It is as though we were trying to tie our shoelaces. That is a fairly difficult job to do while standing. We must first lift one foot and balance ourselves while we tie one lace. Then we do the same for the other foot. If a chair is present, however, we can use its surface by sitting on it, and then the process becomes very simple.

This sort of catalyst action is known as SURFACE CATALYSIS. Naturally, in surface catalysis, the more surface there is, the faster the action. A large chunk of metal has only so much surface. If that same chunk of metal is ground into a fine powder, it has much more surface. It becomes almost all surface, in fact. That's why iron filings and powdered nickel are used as catalysts instead of ball bearings and five-cent pieces.

The enzymes of the body work by means of surface catalysis. That is important to remember.

It is obvious that when a catalyst allows a molecule to make use of its surface it isn't really giving up anything. Once the hydrogen peroxide molecule has rid itself of its extra oxygen atom, what is left releases its hold on the piece of iron surface it was occupying. That patch of surface is then free for another molecule of peroxide. And so on indefinitely.

It is the same principle by which the chair is available for another person once you are through using it while tying your shoelaces. You can imagine a whole line of unlaced people waiting to use one chair. The chair isn't affected and is always ready for the next no matter how many people have already used it. If the people waiting are in no particular hurry, one chair is as good as a million.

For this reason catalysts, and enzymes, particularly, go a long way. The body contains only small quantities of the various enzymes it possesses. Very small quantities. But those small quantities are sufficient.

WHAT ENZYMES ARE

Mankind has been using enzymes since prehistoric times, but it was only a century or so ago that they first realized it.

For instance, if flour is mixed with water to form dough, and the dough is baked, the result is a hard cracker-like material. It is nourishing, but makes for heavy eating. The Bible refers to such a product as "unleavened bread." Nowadays, it is more usually called "matzos." It is still eaten by Jews in celebration of the Passover.

While mankind was still quite primitive, however, it must have been noticed that, occasionally, the dough would puff

HOW AN ENZYME WORKS

ENZYME BEFORE
WORKING

MOLECULE
COMPLETE

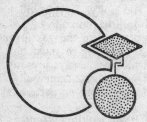

ENZYME WHILE WORKING

ENZYME AFTER
WORKING

MOLECULE
SPLIT APART

up before baking. The result would be that, after baking, the bread was softer, fluffier, and pleasanter to eat. Some prehistoric genius then found out that if a little of the puffed-up dough were saved, it could be added to a new batch of dough and that would puff up also. The little lump of puffed-up dough is called "leaven" in the Bible, and bread made with it is "leavening bread." The ordinary bread we eat today is of this type.

It is now known that leaven works as it does because it contains large numbers of microscopic little plant cells known as yeast. These little yeast cells live on the starch in the dough and their enzymes convert a small part of it to alcohol and to a gas called carbon dioxide. This is the same gas that is used to make carbonated drinks. It is carbon dioxide that fizzes up when we take the cap off a warm soda water bottle.

Since the yeast occurs all through the dough, carbon dioxide forms in tiny bubbles everywhere in it, puffing it up and making it light and fluffy. If you will look at a slice of bread, you will see the millions of holes of all sizes that are present as a result of the gas formation.

Originally, the first batch of dough must have puffed up because yeast cells (which are always present, floating about unseen in the air) got into it accidentally. The use of leaven changed the process from an accidental one to a deliberate one.

Since the yeast is alive, it will multiply if it gets a chance. Thus, if a small piece of leaven is added to a large batch of fresh dough, the yeast will multiply till all the dough is filled with it. The Bible contains a proverb about that: "A little leaven leaveneth the whole lump."

Another age-old use of yeast involves fruit juices. If grape juice, for instance, is allowed to stand under the proper conditions, yeast will grow in it and change the sugar it contains to alcohol. The grape juice has "fermented," in other words, and become wine.

Yeast is very famous for the things its enzymes will do. In fact, the very word "enzyme," which was first used in 1878, comes from two Greek words meaning "in yeast."

Until sixty years ago, it was thought that these yeast enzymes could only work as part of a living cell. In 1897, however, a German chemist named Buchner mashed up yeast cells and found that the resulting juice would work just as well.

Scientists then began to treat yeast juice (or the juice from animal or plant tissues) in various ways in order to concentrate a particular enzyme. It was possible, for instance, to obtain an ounce of liquid that contained as much enzyme as a pound of the original yeast. It was possible to do even better.

Even so, it was difficult for chemists to decide exactly what kind of chemical an enzyme was. The living cell contains so little of any given enzyme, you see, and so many other chemicals are present to help confuse the situation. Even considerable concentration wasn't much help at first. Slowly, however, small pieces of evidence began to pile up to show that enzymes were proteins.

This wasn't proven definitely, however, until 1926, less than thirty years ago. In that year, the American biochemist, Sumner, separated an enzyme named urease from a type of seed called the jack-bean. He obtained this enzyme in the form of pure crystals and collected enough of it to work on. It was then possible to show that one enzyme at least was definitely a protein. Recently, Sumner received a Nobel Prize for this achievement. Since 1926, dozens of other enzymes have been isolated and all, without exception, have been found to be protein.

We can now say that *all enzymes are proteins*.

This makes sense, since enzymes are surface catalysts. The body can construct a protein out of amino-acids (remember the infinite arrangement possibilities we mentioned in the previous chapter) that will have just the proper surface for a certain purpose.

A piece of iron happens to have a surface that suits the hydrogen peroxide molecule *fairly* well. A molecule of catalase, however, is custom-made, so to speak. The hydrogen peroxide molecule fits into its surface perfectly. It is no wonder that catalase is thousands of times as good a catalyst for hydrogen peroxide decomposition as is iron.

It is as though we were to invent a special chair just for shoelace tying—one that would hold our arms and legs in just the correct position to allow us to tie our shoe-laces without reaching.

Thousands of different enzymes are known and each one

is differently designed to suit its own needs. This is an example of how useful it is to have twenty different building blocks out of which to make proteins.

HOW ENZYMES ARE STUDIED

It is not easy to study the enzyme molecule directly. There are too few of them and too many other molecules mixed with them. Instead, biochemists observe their actions. They study the chemical reaction that a particular enzyme catalyzes. Since these reactions exist in the thousands, there are thousands of ingenious methods of enzyme study. We can only briefly describe two or three.

One group of methods for enzyme study involves those reactions which use up gases. For instance, there is an enzyme in potatoes (and some other plants) which catalyzes the reaction of certain chemicals with the oxygen of the air.

In order to study this enzyme, a small piece of potato is cut into very thin slices. These slices are placed in a little glass container and covered with water. Then a little of the proper chemical is added. Instantly, because of the enzyme present in the potato slices, the chemical starts combining with the oxygen in the air above the water. If the container is entirely closed, a partial vacuum is formed inside. If the container is now connected with a fine tube which is dipped in water, the water is sucked part way up the tube because of the vacuum. (The principle is the same as that by which we suck soda up a straw.)

The scientist conducting the experiment carefully notes how far up the tube the water is sucked in a certain time. The higher it is sucked, the more of that particular enzyme is in the potato. In this way, we can obtain quite accurate notions about quantities of enzyme and even about the manner in which it operates. And all the time we're working with amounts far too small to see or weigh.

Naturally, the scientist must be very careful about the conditions of the experiment. For instance, enzymes work with different speeds at different temperatures. So the glass container is dipped into a large tub of water, the temperature of which never varies more than a few *thousandths* of a degree. Also, the scientist must use very pure chemicals, since, as we shall find out later, small quantities of impurities can affect enzymes considerably. He must even add a certain group of chemicals, known as a BUFFER, to keep the enzyme's surroundings just the way they are in the original potato. Every enzyme is adjusted to its own environment and the experimenter must reproduce it as exactly as he can if he wishes to get decent results.

As you see, a scientist has his problems.

Some enzymes catalyze reactions that *produce* gas. When this happens the gas pressure in the little glass container goes up and the water in the tube is pushed downward. Again it is simply a matter of observing how far downward.

One more example. A group of enzymes known as phosphatases cause certain substances to decompose, producing phosphoric acid. Various chemicals can then be added to the container after the enzyme has been working for a period of time. These combine with phosphoric acid to form a blue color. The bluer the color, the more phosphoric acid has been produced, and the more phosphatase you are working with. Very ingenious devices, known as colorimeters, exist. They can measure the strength of a color with great accuracy, so you don't have to depend on your eyes.

Sometimes it is possible to work with pure crystalline enzyme. In that case, you can tell exactly how much enzyme you've used and how much reaction has taken place. It is then possible (by means of calculations we won't go into) to determine how many molecules of a substance have been changed in one minute by a single molecule of enzyme.

Usually, each enzyme molecule can take care of a few thousand molecules of a substance each minute. Our friend, catalase, however, can really go to town. In one minute, a single molecule of catalase can break down *five million* molecules of hydrogen peroxide.

Enzymes are globular proteins. That means their amino-acid chains are woven into complicated patterns that are held together in very delicate fashion. This may sound rather fragile to you. If it does, you are right. Globular proteins are very fragile.

For instance, heat is very hard on proteins. Molecules vibrate as a result of heat. The higher the temperature, the harder they vibrate. It doesn't take much vibration, as you can imagine, to shake loose the complicated structure of a globular protein. And when we say heat, we don't mean a match flame. We don't even mean boiling water. An ordinary warm summer day is hot enough to destroy an enzyme; to cause it to lose its activity, to IN-ACTIVATE it, in other words.

In order to keep enzyme solutions active while they are being studied, it is usually necessary to keep them in the refrigerator. Freezing it solid is even better.

You may ask how enzymes manage to survive in the human body which is always at a temperature of 98.6 degrees Fahrenheit. The answer to that is that they don't really. Enzyme molecules are always breaking down in the body, but the body is always making new ones. In a test-tube at that temperature, the enzyme molecules break down and there is no way of making new ones.

Other conditions can also affect the enzyme's delicate structure. Acids or alkali will ruin them. Small quantities of various chemicals can destroy them. Even as simple a thing as shaking an enzyme solution vigorously may cause the structure of an enzyme molecule to collapse.

All this is of immediate importance to all of us. A living creature depends entirely upon the workings of the various enzymes in his body. If anything interferes with those workings, he can no longer live. Conditions that inactivate enzymes, therefore, will eventually kill human beings.

Usually quite a large number of enzymes must be destroyed in order to kill a man, but sometimes that is not so. There are some enzymes that are of such key importance that any interference with them, even for just a few minutes, is fatal. Small amounts of certain chemicals will do just this, and we call such chemicals POISONS.

We'll discuss one or two of these poisons later in the book, but first we must pause to get some idea of how enzymes go about organizing the body's chemical activities.

SUMMARY

Certain proteins are catalysts. That means they have the ability to make chemicals react with one another very rapidly without themselves being changed. In their absence, the chemical reactions would proceed very slowly indeed. Such catalytic proteins are called enzymes. Enzymes control body chemistry. They are very fragile and easily destroyed. The destruction of even a small number of them may cause illness or even death.

3

Enzymes and Body Chemistry

THE CELL—A BAG OF ENZYMES

EXCEPT FOR THE VIRUSES, all living creatures are composed of cells. Very simple organisms such as yeast and bacteria consist of only one cell apiece. A large organism, such as a human being, contains billions upon trillions of cells. A single drop of blood, for instance, contains about forty billion cells. That's just one drop, and there are thousands of drops of blood in the average man.

As you can imagine, cells are very small. It would take a thousand of even the larger ones to make a line one inch long when placed end to end.

Despite its small size, each cell is a tiny drop of life all by itself. Some cells can live all by themselves, and *do*, as in the case of bacteria. Human cells, however, have lost that ability. They depend on one another so much that they have forgotten how to get along by themselves.

We see the same thing on a larger scale in people. A primitive man could live alone and survive. A man who has lived all his life in New York City wouldn't live long, probably, if he suddenly found himself on an uninhabited island. He wouldn't know how to find food and shelter, how to protect himself against wild beasts, and so on. He would be too accustomed to living as part of a large group. And so it is with human cells.

Groups of cells, taken all together, are more advanced than single cells, even if the latter are more independent. The human being is more complex and advanced than a germ, just as a modern city is more complex and advanced than a caveman's cave.

The living matter inside a cell is called PROTOPLASM. The protoplasm is divided into two parts. Near the center of the cell is a part which is denser and thicker than the rest of the cell. It is NUCLEUS. The rest of the cell is CYTOPLASM.

Like any other living thing, cells grow and multiply. Most cells multipy by dividing down the middle. Then there are two cells where only one existed a moment before. The cell nucleus is in charge of seeing that cell division takes place properly. The cytoplasm takes care of the day-by-day life of the cell.

Cells in different parts of the body vary in their shape according to the work they must do. Fat cells are just tiny blobs of fat surrounded by a thin layer of protoplasm. The red cells of the blood are little disks that contain a protein called hemoglobin, which carries oxygen to all other cells of the body. Red blood cells are so simple they don't even have a nucleus and so cannot grow or divide. They have to be manufactured fresh continually in the bone marrow.

Nerve cells have irregular shapes with long thread-like fibers sticking out of them. Impulses and sensations travel along those fibers. Muscle cells are long and thin. They can contract into short, thick cells whenever necessary.

Some cells are so specialized that they have abandoned almost everything but their main function. They have even lost the ability to multiply. A baby is born with all the brain cells, for instance, that it will ever have.

Still other cells are always growing. The cells of the skin grow and divide throughout life. As old skin wears off of the body, new skin is always formed to take its place.

Now how does each cell do its job? Perhaps you may have guessed from what we said in the previous chapter that each performs its functions by means of the enzymes it contains. Every different kind of cell has its own special group of enzymes. Some enzymes can be found in almost all cells, some only in a few cells. The point is that each cell has the kind it needs and no others.

You may wonder whether there is room in the cell for all the enzymes it needs. After all, the cell is a very small object.

Well, let us consider the liver cell. The liver is the handy-man organ of the body. It has more different chemical jobs to do than does any other part of the body. Therefore it would need more different kinds of enzymes.

Fortunately, the liver cell is big enough to hold more than *two hundred trillion* (that is, 200,000,000,000,-000) molecules. Ninety-eight per cent of them, to be sure, are just water molecules. However, there are still 50,000-000,000 (fifty billion) protein molecules in a single liver cell. That's quite a number. Even if we suppose that only one protein molecule in a thousand is an enzyme molecule, the liver cell still has 50,000,000 enzyme molecules. And if it takes one thousand molecules of each particular enzyme to do the job, there is still room for over 50,000 *different* enzymes.

So you see, cells are quite big enough to get along. You can think of each cell as a tiny bag of enzymes; a tiny bag of a great many enzymes.

HOW ENZYMES KEEP ORDER

Molecules are always changing. They break in two or more pieces. Or they may add on an atom or two. Or they may exchange atoms with another molecule. All sorts of changes are going on all the time.

Inside the cell especially, there could easily be great confusion. Large numbers of all kinds of molecules are roving about and bumping into one another. Anything could happen, but doesn't. Only certain things happen.

Why is that? Well, to begin with, it seems that although some chemical reactions are fast, others are slow. When small pieces of copper are dropped into a strong acid known as nitric acid, there is a fast chemical reaction. The copper turns green and dissolves, while a brownish gas is formed. On the other hand, the rusting of iron in damp air is a slow reaction. It might take days before enough rust accumulates to be visible.

Some reactions are slower still. Do you realize that paper is burning all the time—the very paper you're looking at right now? The process is so slow that there is no flame and no heat, but it's proceeding just the same. If someone were to pick up this book a hundred years from now, he would find the pages yellow and brittle. They would be half-burnt, really. Eventually, with the years, the pages will completely disintegrate. There would be nothing left but ash. Of course, the process could be hastened easily enough. Raise the temperature of the paper by applying a match flame and it's all over in a second.

Most of the important reactions inside the cell are of the slow type; even of the very slow type. To show why that is important, let us suppose we have a number of molecules of a certain kind moving around inside a cell. A hundred thousand of them, perhaps. Let's call them Molecule A.

Molecule A may do any one of a number of things. It may break in two. It may exchange atoms with Molecule B, or with Molecule C, D, E, F, all the way to Z and beyond. It may simply rearrange the atoms within itself and become a new molecule without any outside help at all. Different samples of Molecule A might be doing all these things simultaneously. The point is that every one of these possible reactions is a slow one. It might take days before even a few of the hundred thousand molecules of Molecule A could manage to go through any one of them.

But also present in the cell is a certain enzyme which catalyzes the reaction of molecule A with molecule F. That *one* reaction is hastened. All the other reactions are not affected. They still continue at their usual slow rate.

The reaction of Molecule A with Molecule F is *so* hastened, however, that all one hundred thousand molecules of Molecule A react with Molecule F in a second or so. There is just no time for any of Molecule A to have anything else happen to it. In this way, the enzyme creates order out of what might be disorder.

One of the virtues of enzymes in general is that they are particular about what they do. In the case we have just mentioned the enzyme catalyzes the reaction of Molecules A and F, *and nothing else*. This is what we mean when we say an enzyme is SPECIFIC.

Of course, there are sometimes limits to this specificity. If another molecule were substituted for Molecule F,

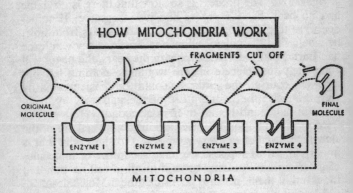

which was *almost* the same but not quite, the enzyme might not know the difference. It might catalyze a reaction of Molecule A with the substitute. Some enzymes are so specific that they can't be fooled by any substitution, however close. Other enzymes are a little more tolerant.

Still even the most tolerant enzyme is much more specific than other types of catalysts. Such catalysts as nickel, platinum, iron and water will catalyze many individual reactions. If cells were forced to rely on such catalysts they could never be certain that only the reactions they wanted were happening. So chalk up another point for the useful protein molecule.

Of course, if the cell needed to have one type of molecule react in two or more different ways, it could simply manufacture two or more different enzymes. If it wanted more of one reaction than the other, it could make more of that particular enzyme. Also, various enzymes can be so arranged within the cell as to catalyze different reactions at different times.

The cell has little MITOCHONDRIA which hold enzymes arranged in order. These resemble factories. The different molecules are the materials being worked with and the enzymes are the workmen. Each workman is in his place and each workman has his one particular job.

WHY ENZYMES WORK BOTH WAYS

Until now, we've been more or less bragging about what enzymes can do, how useful they are and so on. It is important to show the other side of the picture as well. Enzymes have very definite limitations, too.

Here's one of them. An enzyme can't make a reaction go, unless that same reaction would go (even though very, very slowly) *without the enzyme*.

If a particular reaction didn't go at all, then just adding an enzyme couldn't make it go.

To get the picture a bit more clearly, let us suppose we're sitting in a parked car. The car is on a hill, but the brakes are on. It isn't moving at all as far as we can see though probably it is slipping downhill *very* slowly. Now release the brakes. The car starts rolling and begins moving faster and faster. If the car were well-greased and the pavement were very smooth, this would happen even if the slope of the hill were so gentle it was invisible to the eye.

But if the car were parked on an absolutely level stretch of ground, so that there were no slope at all, releasing the brakes wouldn't *make* it move. And certainly, releasing a car's brakes would never make it roll *up*hill.

Well, when two chemicals react with one another, they lose something that scientists call FREE ENERGY. This loss of free energy is what makes them react. It means they have moved into a more stable position by reacting. In the same way, a car moves into a more stable position when it loses gravitational energy by rolling to the bottom of the hill. (Once the car gets to the bottom of the hill, you can leave the brakes off. It won't roll in either direction. That's what we mean by saying the position has become more stable.)

In fact, you can just picture any chemical reaction as a process whereby chemicals roll down an "energy hill."

When the reaction is finished, the chemicals are at the bottom of the hill. When two chemicals react very slowly, it is because there is something about their molecules that prevents them from rolling freely down the energy hill. It's as though they had their brakes on. The action of the enzyme is to release those brakes and let them roll.

When there is no energy hill at all, the reaction won't proceed. And releasing the brakes by adding an enzyme doesn't help, any more than it would in the case of an

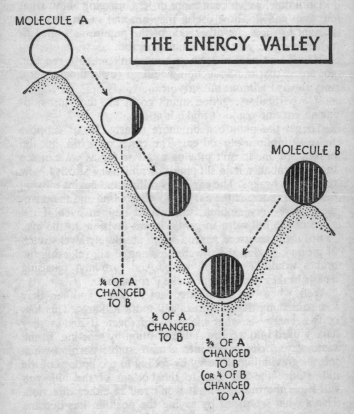

MOLECULE A

THE ENERGY VALLEY

MOLECULE B

¼ OF A CHANGED TO B

½ OF A CHANGED TO B

¾ OF A CHANGED TO B (OR ¼ OF B CHANGED TO A)

automobile on level ground. That's what we mean when we say an enzyme can't make a reaction go, unless that same reaction would go without the enzyme.

Now we're ready for an important point. When Molecule A reacts so as to change to Molecule B, it often doesn't make the change completely. As it changes to Molecule B, it is rolling down the energy hill all right, but when it comes to the bottom, perhaps only three-quarters of the molecules have changed. To change more of its molecules, it would have to start rolling *up* a neighboring hill, *and it can't.*

Molecule B is up on top of that neighboring hill and if it were to change to Molecule A, it would roll down that hill. But when it came to the bottom of the hill, it would be only one-quarter changed. No matter which hilltop you start from, you see, you would end up in the same place—at the bottom of the hill, or, rather in the valley between two hills. Each time, three-quarters of the molecules would be Molecule B, and the rest would be Molecule A.

Now an enzyme can't change the facts of life as far as that energy valley is concerned. It can make Molecule A change to Molecule B faster, but it can't make the change move past the bottom of the valley.

And here's something curious. Enzymes are two-faced. The same enzyme that can change Molecule A to Molecule B *can also change Molecule B to Molecule A*. But it can only do this when it's on the Molecule B hill and it can't make Molecule B move past the bottom of the valley either.

HOW ENZYMES ARE PERPETUATED

Earlier in this chapter we said that the nucleus of the cell was in charge of cell division. Unfortunately, most of the details of the process are as yet unknown. Still we can describe a bit of it.

Inside the nucleus are small patches that can react with certain dyes to become strongly colored. Biologists noticed them for that reason and called the material in the patches CHROMATIN from the Greek word for color.

In the process of cell division, the chromatin collects

into little rods of varying size. The rods are called CHRO-
MOSOMES. In the nuclei of human cells there are forty-six
such chromosomes, existing in pairs. There are twenty-
three *pairs* of chromosomes, in other words. Each kind of
creature has its own fixed number of chromosomes. A rat
has thirty-eight chromosomes, a grasshopper twenty-four
and a housefly only twelve. A crayfish, on the other hand,
has over two hundred chromosomes.

Before a cell divides, every chromosome lines up in the
center of the cell and splits in two. The two halves of each
chromosome move apart and when the cell does divide,
each new cell has a duplicate of all the original chromosomes.

It is these chromosomes that control a cell's characteristics. A cell's nature is determined by the kind of chromosomes it has. That is why the chromosomes are so
carefully split and divided. By having each new cell get

HOW CHROMOSOMES SPLIT UP AND RE-UNITE

FATHER CELL

MOTHER CELL

SPERM

OVA

FERTILIZED OVUM

exactly the same share, it is made certain that the two cells are similar to each other and to the parent cell.

Most plants and animals (the human being included) produce special cells out of which new individuals altogether are formed. These are called SEX CELLS. The sex cell produced by a female is called an OVUM, and the sex cell produced by a male is a SPERMATOZOON. Sex cells contain only half the normal number of chromosomes. Human sex cells, for instance, have only twenty-three chromosomes, one for each pair. The first step, however, in the growth of a new individual is for a spermatozoon to join an ovum and combine nuclei. The resultant FERTILIZED OVUM now has forty-six chromosomes once again. Twenty-three have been inherited from the father and twenty-three from the mother.

In each generation there is thus a mixing and remixing of chromosomes and no two individuals are ever alike, unless they are "identical twins." In the case of identical twins, the fertilized ovum splits in two and begins the growth of two individuals. Both halves of the ovum have identical chromosomes and so the two babies look very much alike, and develop in the same way. Since chromosomes also decide the sex of the baby, identical twins are always both girls or both boys.

"Fraternal twins" are due to the formation of two different ova at the same time. They naturally have different sets of chromosomes. Fraternal twins don't have to look any more alike than any other pair of brothers and sisters. They don't even have to be of the same sex.

Every chromosome is actually a chain of protein molecules which are called GENES. Genes are strung along a chromosome as pearls are on a necklace. The genes have a certain chemical resemblance to viruses. In fact, some people consider them a sort of tame, house-broken virus. According to this view, genes would compare to viruses as watch-dogs compare to wolves.

Each gene is thought to be in control of a single characteristic of an organism. For instance, there is a gene for blue eyes and one for brown eyes; one for straight hair and one for wavy hair. Every human being has thousands of different genes scattered through his various chromosomes. Whenever a chromosome splits in two, during cell division, each gene duplicates itself exactly and both daughter cells get one apiece.

How does a gene control a particular characteristic? It is now thought by many people that each gene is in charge of manufacturing one particular enzyme in the cell. And now finally we are back to enzymes and you can see why we have taken this small detour among the chromosomes.

But how does a gene manufacture an enzyme? For that matter, how does a gene duplicate itself? This is probably the most important unanswered question in biochemistry today.

There are theories, of course. There are enzymes that take proteins apart and separate them into amino-acids. But you'll remember from the preceding section that enzymes can work both ways. These protein-splitters can also put amino-acids back together again.

Apparently, then, the beef protein we eat (or milk protein, or wheat protein) is separated into amino-acids and then put together in a different arrangement to make human protein. But how is the arrangement figured out, when there are so many possibilities?

Here is where the gene comes in. Genes are nucleo-proteins. The non-protein part of the molecule is the nucleic acid I mentioned in Chapter 1. Each gene contains its own variety of nucleic acid. Each different nucleic acid somehow acts as a model for the formation of a particular enzyme. Nucleic acids, therefore, control amino-acid arrangements.

How? Chemists just began working out the method in the 1950's. The nucleic acid of the chromosomes forms a "messenger" molecule which leaves the nucleus and joins particles in the cytoplasm which are called RIBOSOMES.

In the ribosomes are tiny fragments of nucleic acid molecules. There are a number of kinds of these fragments and each will attach its own particular type of amino-acid. These nucleic acid fragments carry their amino-acids to the "messenger" molecule and use its structure as a guide. They line up to match the structure and each transfers its amino-acid. In this way, an entire protein molecule is formed with an exact structure according to the original design of the chromosome's nucleic acid.

You may wonder how enzymes can control characteristics. How can they decide blue or brown eyes, for instance? Well, eye-color is due to a pigment called melanin. When the eyes contain very little melanin, they appear blue. With more melanin, they are brown. Melanin is formed in the body as a result of a chemical reaction which is catalyzed by the enzyme, tyrosinase. The amount of melanin formed depends upon the amount of tyrosinase present. Possession of a gene producing much tyrosinase will result in brown eyes. A gene that produces less tyrosinase makes for blue eyes.

What happens when a cell splits in two without proper duplication of genes? Sometimes the daughter cells just can't live. At other times, the cells survive, but with a changed chemistry. Some biochemists think that cancer cells may originate as the result of such imperfect duplications. Actually, considering all the times that genes duplicate themselves, such accidents happen very, very rarely.

When a gene duplication goes wrong in a germ cell, the result is a MUTATION. A mutated offspring is one which is quite different from either parent in one way or another. Either it is missing a gene or else it has a new gene altogether. For instance, suppose the gene which manufactures tyrosinase slips up while duplicating and ruins itself completely. The offspring can make no melanin whatever. His skin is very fair. He cannot tan. His hair is white. His eyes have no color except for the redness of little blood vessels. Such people are called ALBINOS. This is an illustration of a mutation. They occur every once in a while because of the slip-up of a *single gene*.

Most mutations are for the worse. Occasionally, though, useful ones may occur. A larger, keener brain may develop, perhaps, or a more flexible hand with a longer thumb. It is as a result of the occasional useful mutations that EVOLUTION takes place. It was a series of mutations that changed the primitive ape-men to the human beings we now are.

SUMMARY

Each different enzyme in the body controls one and only one particular reaction. Therefore the working of a particular cell depends on the kind of enzymes it contains. Enzymes are formed under the direction of genes. Each gene is responsible for one enzyme, probably. Whenever a cell divides and becomes two cells, the genes also divide so that each new cell has its share. When organisms give birth to young, the cells of the offspring obtain half their genes from the mother and half from the father.

4

Enzymes and Digestion

THE FOOD WE EAT

THE RAW MATERIAL of life is food. For human beings, food consists usually of material that was once alive or was part of something that was once alive. We drink water and eat salt, of course, but those are about the only exceptions to the rule.

Since food is derived from living creatures, it is made up of the same type of substances as we are. First, and usually foremost, there is water. Some foods are very watery indeed. Leafy vegetables are nine-tenths water. Even a "dry" food such as bread is one-third water. (*Really* dry bread is quite unpleasant to eat.)

Secondly, food contains small quantities of minerals and vitamins. There will be special chapters for them later on.

The rest of the food consists of three kinds of materials. In addition to proteins, they contain FATS and CARBO-HYDRATES. The proteins provide the amino-acid building blocks of living tissue, as we have explained earlier in the book. The fats and carbohydrates provide energy for the body.

If we were to compare the body to an automobile, ordinary proteins would be the frame and the motor; the enzymes would be the controls; and fats and carbohydrates would be the gasoline.

Most foods in their natural state contain some of all three of these substances. However, man can refine food, so that some of the commonest things we eat are not at all balanced. For instance, table sugar is pure carbohydrate. Butter, lard, and olive oil are almost pure fat. Some kinds of cheese contain scarcely anything but protein and water.

There are two main varieties of carbohydrates: SUGARS and STARCHES. The sugars are much simpler than the starches. The molecular weights of the sugars, that is, the sum of the weights of all the atoms in the molecule, are usually below 400. There are some particularly small sugars whose molecular weight is about half that. Three of these simple sugars are important in the diet. These three are GLUCOSE, FRUCTOSE, and GALACTOSE. They are quite similar to one another chemically.

These sugars rarely occur by themselves. Honey contains a mixture of glucose and fructose and so do some fruits. Those are about the only places they are found. In fact, glucose is sometimes called "grape sugar" and fructose is called "fruit sugar."

Then what makes them important in the diet? It seems they occur in combination. For instance, glucose and fructose can be hooked together, to form a single molecule. The result is a "double sugar" called SUCROSE. It is sucrose that is the sugar we use to sweeten coffee. Cane sugar and beet sugar are both sucrose. So is maple sugar.

It may be hard to realize that until just a few centuries ago, sugar, as a substance by itself, was unknown. In the Middle Ages, the only way in which Europeans could sweeten food was to add honey to it.

Another form of a "double sugar" molecule arises when glucose and galactose are hooked together. The result is LACTOSE, or "milk sugar." As the name implies, lactose is found in milk. What is more, it is found nowhere else.

Sugars are not all equally sweet. The sweetest sugar is fructose, and next is sucrose. Glucose is only moderately sweet and lactose is almost tasteless. (That's why natural milk is only very slightly sweet even though it contains quite a bit of sugar.) All these sugars are equally useful as sources of energy, however, whether they are sweet or not.

HOW DOUBLE SUGARS ARE BUILT UP

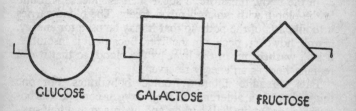

GLUCOSE GALACTOSE FRUCTOSE

SUCROSE (ORDINARY SUGAR)

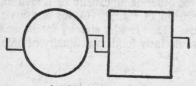

LACTOSE (MILK SUGAR)

Starches, as we said in the first chapter, are very large molecules consisting of chains of many glucose molecules hooked together. The main difference between the properties of starches and sugars is that sugars mix freely with water. That is what we mean when we say sugar is SOLUBLE. We observe this fact every time that we add sugar to coffee. Starches, on the other hand, do not dissolve in water. They are INSOLUBLE.

In the body, therefore, a sugar such as glucose is found in the blood, with which it mixes freely. The blood carries it to all parts of the body so that it may be used for energy. Starch, however, occurs in the liver. Because it is insoluble, it isn't washed out of the liver by the blood, so that it can remain there as an energy reserve.

Fats are quite different from carbohydrates. For one thing, there is a difference in their oxygen content. Carbohydrates are about half carbon and hydrogen and half oxygen. Fats, on the other hand, are nine-tenths carbon and hydrogen and only one-tenth oxygen. Now it is carbon and hydrogen that are the sources of energy. It follows that fats contain more *concentrated* energy than do carbohydrates.

Animals, for that reason, store their food mainly in the form of fat. In the human being, for instance, the total amount of starch stored in the liver contains only enough energy to keep the body going for about fifteen hours. Enough fat can be stored, however (and sometimes is), to last a human being for months.

Plants, on the other hand, usually keep their main energy supply in the form of starch. (Olives and avocadoes are two exceptions.) That is why starchy foods are usually of plant origin while fatty foods are usually of animal origin.

SMALL MOLECULES OUT OF LARGE

Here we arrive at a hitch in the proceedings. All these substances in food are of no use to the human body the way they are. No use at all. The proteins, fats, and starch that make up our body are quite different from the proteins, fat, and starch of the food we eat. Even the double sugar

HOW FOOD PROTEIN IS CHANGED TO HUMAN PROTEIN

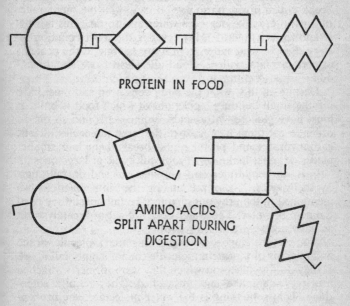

PROTEIN IN FOOD

AMINO-ACIDS
SPLIT APART DURING
DIGESTION

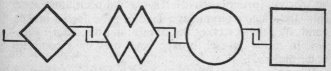

AMINO-ACIDS REARRANGED AND COMBINED
TO FORM HUMAN PROTEIN

molecules are no good to us directly. Our bodies never contain sucrose, and only rarely contain lactose.

In fact, food molecules, particularly protein, can be downright deadly. If a strange protein (even one from another human being, sometimes) gets into too close contact with us, we may become "sensitized" to it. After that, just breathing the protein or touching it can cause us to itch, break out in hives, or sneeze. It can swell the membranes of our nose, make our eyes water and so on. This is what is known as ALLERGY. Hay fever is the most common allergy. People who suffer from it are allergic to the proteins of certain plant pollens. Food allergies of one sort or another are also common.

Despite all this, what the body does need and must have are the small building blocks out of which food is built. It must have glucose, fatty acids, amino-acids and so on. In order to get them it must take the large molecules of fats, carbohydrates and proteins and break them into smaller pieces. It must make small molecules out of large ones.

Fats, carbohydrates and proteins can and do fall apart by themselves. These are among the slow reactions we mentioned in the previous chapter. To fall apart, they need the help of water. The water does not act as a catalyst but actually takes part in the reaction.

You see, the glucose portions of a starch molecule, or the amino-acids of a protein molecule cannot simply fall apart. If they do, the places at which they were formerly attached are left incomplete and unsatisfied. They require something. That something is the water molecule. One broken-off glucose (or amino-acid) takes a hydrogen atom away from the water molecule as a kind of salve for its broken bond. Its neighbor takes the oxygen and the other hydrogen. In this way, each broken bond uses up a complete water molecule.

When the atoms of water patch up two pieces of a broken molecule in this way, the process is known as HYDROLYSIS. Hydrolysis is one of the most common reactions in the human body. That is one of the reasons why the body contains so much water. (Another reason is that when chemicals are intimately mixed with water they react with one another more quickly and easily than if they existed as solid chunks.)

Fats, proteins, and carbohydrates—all three—can be reduced by hydrolysis to the simpler building-blocks that the body needs. The only problem is to speed up the process. One way of doing this is to use acid. In one place, as we shall shortly see, the body actually does this. In general, however, the body has a better solution. This, as you have probably guessed, involves the use of enzymes.

Enzymes which catalyze hydrolysis reactions are called HYDROLYZING ENZYMES. When food-stuffs are hydrolyzed in the body by means of such enzymes, the process is known as DIGESTION.

Just as there are three types of food components there are three types of digestive enzymes. Those enzymes which specialize in hydrolyzing starches are called AMYLASES. Those that hydrolyze fats are LIPASES. Those that hydrolyze proteins are PROTEASES.

(Names of enzymes almost always end in "ase." This is because chemists have gotten together in the past and agreed to make an "ase" ending a sort of trademark for enzymes.)

ASSEMBLY-LINE
TECHNIQUES IN THE BODY

The food canal is a long tube that starts at the mouth, goes through the body, and ends at the anus, the opening through which wastes are passed. Since it is about thirty feet long in the adult, it must be coiled up if it is to fit inside the body. It has wide parts and narrow parts and every part has a particular function.

Into various portions of this canal, watery solutions pour more or less continually. These SECRETIONS usually contain digestive enzymes and are manufactured by organs (some large, some small) that are called GLANDS.

There are SALIVARY GLANDS, for instance, in the cheeks and under the tongue, which continually manufacture saliva and pour it into the mouth. Saliva contains a starch-splitting enzyme called SALIVARY AMYLASE. Saliva's job is not only that of moistening the food during chewing, so that

HOW HYDROLYSIS WORKS

PART OF A STARCH MOLECULE

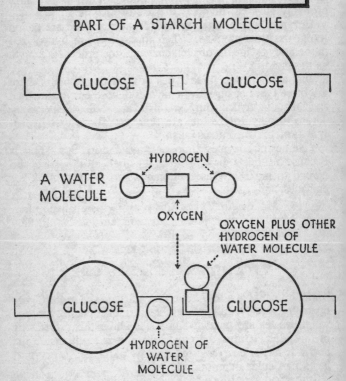

GLUCOSE MOLECULES AFTER HYDROLYSIS

it becomes easier to swallow. It also starts the process of digestion.

After being swallowed, the food passes through a narrow tube called the ESOPHAGUS into the STOMACH. The stomach is one of the wide portions of the food canal. It has muscular walls, which help squeeze the food and mix it well with digestive secretions. (When the stomach is empty, the writhing of the walls gives rise to "hunger pains.") In the stomach lining are numerous microscopic glands which secrete GASTRIC JUICE.

This gastric juice is remarkable in that it contains hydrochloric acid, the strongest acid in the body. The stomach is the one place where the body uses strong acid to speed up hydrolysis. Still, even here the main work is done by a protein-splitting enzyme, called PEPSIN, which is also present in the gastric juice.

Pepsin is an example of an enzyme whose name does not end in "ase." It seems that the digestive enzymes are manufactured in quantity in the body and, what's more, are not confined within the cell but are poured into the stomach and intestines. For this reason, they can be more easily obtained and studied than most enzymes. And they have been known longer. In fact, they have been known for so long that they were all given names before chemists could get together and decide on the "ase" ending. Some of them, like pepsin, still retain their old-fashioned name.

When the food leaves the stomach, it enters the SMALL INTESTINE, where the main business of digestion takes place. The small intestine is twenty feet long and is coiled up until it fills a person's abdomen almost completely. At its very start, two large glands, the LIVER and the PANCREAS, pour their secretions into it. The pancreatic juice contains a number of enzymes. One of them is a protein-splitter and is called TRYPSIN. There is also a fat-splitter, PANCREATIC LIPASE, and a starch-splitter, PANCREATIC AMYLASE.

In addition, there are microscopic glands in the lining of the small intestine that secrete intestinal juice, which contains still other enzymes.

Finally, the food moves into the LARGE INTESTINE. No digestion takes place there. Instead, some of the water which the various glands have poured forth into the food canal is taken back into the tissues where it is needed. What is left of the food is then eliminated from the body.

Now why do we need such a long, complicated digestive tube? Why not just a stomach, for instance where all the digestion could take place once and for all? The answer to these questions is that the body finds it more effective to run things on an assembly-line basis.

In automobile factories, the various parts of an automobile pass down a moving belt along which different workers stand. Each worker has one particular job to do over and over again as the parts pass by. By the time the parts reach the end of the belt, they have been put together into a complete car. If all the workers tried to do all their jobs in the same place and at the same time, there would simply be endless confusion and little accomplished.

So it is with the food canal. The different substances in food pass down the digestive tube and enzyme after enzyme gets a whack at them. Each enzyme does its own particular job and takes its proper turn.

Suppose we take carbohydrates, for instance. As food is chewed in the mouth, salivary amylase starts to catalyze the hydrolysis of the starch it contains. It breaks the long chains of glucose into somewhat shorter chains. These shorter chains are called DEXTRINS. In the stomach, the hydrochloric acid breaks up the dextrins into still smaller pieces.

As soon as the food enters the small intestine, the pancreatic amylase tackles what is left of the dextrins and breaks it up into very short chains of only two glucoses apiece. This double glucose molecule is called MALTOSE. The process of digestion is not finished even then. Further down in the small intestine, there is the intestinal juice, which contains the enzyme MALTASE. Maltase has the special task of breaking maltose in two, and making glucose out of it.

Then, and only then, can the body get any use out of starch. It is only when starch has been broken all the way down to single glucose molecules, bit by bit, that it can be absorbed into the body.

The intestinal juice also contains SUCRASE and LACTASE. The former splits sucrose into glucose and fructose. The latter splits lactose into glucose and galactose. Fructose and galactose can be absorbed into the body just as glucose can.

The situation is similar in the case of proteins. The two chief proteases, pepsin and trypsin, attack the amino-acid chain in different places. First the pepsin tackles the protein molecule in the stomach, breaking the chain wherever it can. Then the trypsin breaks it in new places, once the food reaches the small intestine.

When both enzymes are through with their separate jobs, all that is left of the protein molecule are short amino-acid chains called PEPTIDES. They contain only two, three or four amino-acids apiece.

In the intestinal juice there are various enzymes called PEPTIDASES. These are designed for the precise purpose of handling these peptides. Almost every type of peptide has a peptidase all its own. The result is that the peptides are broken down into single amino-acids. These, finally, are absorbed into the body.

Fats are digested in a single step by pancreatic lipase. However, there are special difficulties here. Fats are very insoluble in water. (There is even a well-known proverb stating this: Oil and water don't mix. Oil is simply liquid fat, of course.) The result is that the enzyme (which is dissolved in water) has trouble getting at the fat.

Fortunately, liver secretion (which is called BILE), although it contains no enzymes, does contain certain chemicals which help out. These are the BILE SALTS. They have a soap-like action and help break the fats into tiny droplets. These fat droplets mix well with the water and digestion proceeds.

We have a familiar example of this sort of procedure in our own kitchens. In ordinary milk, the cream (which is the fatty part of milk) separates from the watery part and floats on top. This can be prevented by dividing the cream into very small droplets. The milk, in other words, is "homogenized." In such homogenized milk, the cream mixes well with the water and no longer separates. Well, the same thing happens to fat in the small intestine, with the help of the bile.

WHEN IS
A PROTEIN A GOOD PROTEIN?

The proteins, fats, and carbohydrates from meat, milk, vegetables and any other articles of diet are all broken down to the same simple substances in the process of digestion. Why is it, then, that all foods aren't equally "good for you"? There are several answers to this.

In the first place, some components of foods can't be digested at all. An important case is the carbohydrate, cellulose (plus some of its relatives). Paper and cotton are almost pure cellulose. The molecule of cellulose is made up of glucose molecules which are hooked together in such a way that no human enzyme can tackle it. That's why we can't get nourishment by eating paper and cotton.

Many fruits and leafy vegetables contain a considerable percentage of this indigestible cellulose. They are poor energy sources, when compared to meat or grain. (However, they are valuable for their minerals and vitamins, so don't stop eating them.)

Animals such as cows, sheep and goats live on grass, which contains a great deal of cellulose. Yet they have no cellulose-splitting enzymes either. Fortunately for themselves, though, they have special stomachs where the grass is stored a long time. In these stomachs, certain bacteria also grow and these bacteria *do* have cellulose-splitting enzymes. Of the glucose that is formed, the bacteria take some and the animal takes the rest. It is a happy partnership and biologists call it a SYMBIOSIS. As we shall see in a later chapter, human beings also have partnerships with the bacteria in their intestines, but for a different purpose.

Secondly, the different food substances are of varying importance to the diet. Suppose the food you ate contained very little fat. That wouldn't bother your body a bit. It would take the carbohydrate you eat and turn it into fat. It happens all the time. Everyone knows what starchy foods will do for the waistline.

If both fats and carbohydrates are low, the body is still

WHEN IS A PROTEIN A GOOD PROTEIN? 63

not at a loss. It can manufacture both out of the proteins of the diet.

Where the body *does* get stuck is in the case of a shortage of proteins. It cannot manufacture proteins out of fats and carbohydrates. Proteins require nitrogen, and neither fats nor carbohydrates have any. So proteins can only be obtained for the body by making certain that protein is in the food. It is impossible to live on a diet of starch, butter and sugar. You can get all the energy you need, but you can't build tissue.

Meat, milk, eggs, beans, nuts and grains are good protein sources.

Finally, there are differences between one protein and another. Of the twenty different amino-acids, some can be formed by the human body out of other amino-acids. This means that if there is a shortage of that particular amino-acid in the food, it can be made up for. Another amino-acid, of which there may be more than enough, is partially converted and everybody is happy.

But *eight of the amino-acids cannot be formed from other amino-acids*. If the body is short of any of those eight, it is out of luck. It must wait until it finds them in the food. It cannot manufacture them.

These eight amino-acids are the ESSENTIAL AMINO-ACIDS. Any protein that does not contain ample quantities of each of these eight essential amino-acids is an INCOMPLETE PROTEIN. The proteins of gelatin and of corn are examples of such incomplete proteins. White rats, when fed on artificial diets in which only incomplete proteins are given them, lose weight and die.

In general, animal proteins contain more of the essential amino-acids than do plant proteins. The best proteins of all are in milk and eggs.

In addition to all this, a diet must contain certain minerals and vitamins in order to be satisfactory, but this is a subject for another chapter later on in the book.

SUMMARY

Digestion is the process by which the molecules of the food we eat are broken down into smaller molecules which are then absorbed into the body and built up into the material that composes our own tissue. The body forms different enzymes for this purpose in the digestive glands which discharge their juices into the stomach and intestines. Different enzymes do different parts of the job. There are starch-splitting enzymes, fat-splitting enzymes, and protein-splitting enzymes. As the food passes along the intestine, each different kind of enzyme works upon it in turn. By breaking food into smaller molecules and building it up again, the body can make fat out of starch and protein and it can make starch out of fat and protein. To make protein it must have a few of the amino-acids present in the diet. These are called the essential amino-acids. There are eight of them. The other amino-acids can be manufactured by the body out of the essential eight.

5

Enzymes and Energy

THE AIR WE BREATHE—AND WHY

THE BODY USES fats and sugars as a source of energy. In getting the energy it needs out of these molecules, it makes use of a special kind of enzyme. However, before we can talk about these enzymes, we should spend a little time on explaining a few things about energy itself.

Energy is something that makes it possible to do work. You are using up energy when you move your arms and legs. Your heart and lungs are continually using up energy, night and day. Your body uses up energy in just keeping warm. Every day the average man uses up enough energy to bring at least thirty-five quarts of ice water to a boil.

Where does all the energy come from?

If we were to ask that same question of the nation's factories instead of our bodies, the answer would be simple. Our industries use tremendous quantities of energy. One important way in which factories get the energy they need is by burning coal.

Coal is mostly carbon. When coal burns, the carbon atoms combine with the oxygen in the air. They form the gas, CARBON DIOXIDE. A molecule of carbon dioxide is made up of one carbon atom and two oxygen atoms. When carbon reacts with oxygen, energy is given off. If we watch coal burning, we can see the energy in the form of light, and feel it as heat. Light and heat are two of the most familiar ways in which energy shows itself, but they are not the only ways by any means.

Another element which combines with oxygen to produce energy in the form of light and heat is hydrogen. This element produces water when it combines with oxygen. (You will remember that the water molecule is made up of two hydrogen atoms and one oxygen.) Sometimes hydrogen explodes and then energy is displayed in a new form. An explosion moves things—sometimes very violently. When objects are moving, they possess KINETIC ENERGY.

Carbon and hydrogen can combine with oxygen to produce energy even when they form part of a molecule. For instance, gasoline molecules contain both carbon and hydrogen. Gasoline will combine with oxygen, or "burn," as the common expression is. In burning, it will produce *both* carbon dioxide and water. Like carbon and hydrogen, gasoline will give off heat and light when burning, and like hydrogen, it can explode.

Wood and paper have molecules which contain carbon, hydrogen and oxygen and they will burn also to give heat and light.

Our body obtains energy in just the way our industries do. It combines carbon and hydrogen with oxygen. The carbon and hydrogen it uses occur in the molecules of the fats and carbohydrates in our body.

To be sure, the body doesn't burn fats and carbohydrates the way a furnace burns coal. There is no flame in the body and no light. Instead, the molecules of fats and carbohydrates combine with oxygen (that is, OXIDIZE) very slowly. There is an even and steady flow of heat and certain other forms of energy as a result.

Two things are needed, therefore, for body energy. First, the fuel (which can be either fats or carbohydrates). Secondly oxygen, which occurs all about us, and over us as well, since it forms one-fifth of the air we breathe.

The body has ways of storing fats and carbohydrates so that it is possible for us to go without eating for days and weeks. Our body has no way of storing oxygen, however.

Since oxygen is always around us, the body has never felt the need for developing ways of storing oxygen. Since we are using energy every minute of the day and night, we need oxygen all the time. Because we cannot store oxygen, shutting off the windpipe for five minutes can be fatal.

The air extends upward for a hundred miles or more, but it gets thin very rapidly as we move upward. Only the bottom two miles is thick enough to live in for any length of time.

Sixteen times every minute, whether we wake or sleep, we inhale and suck air into our lungs, then exhale and push air out of them. The air we inhale, if it is normal fresh air, is 21 per cent oxygen. The air we exhale is only 16 per cent oxygen. The missing 5 per cent has been absorbed by the body during the time the air was in our lungs.

When we are exercising or working hard, we need extra oxygen to supply extra energy. Our breathing speeds up. We "pant."

The fat and carbohydrate, in combining with oxygen, produces water and carbon dioxide just as gasoline would. The water shows up in the form of vapor. On cold days, the water vapor turns into droplets of liquid water. These droplets form a visible cloud so that our breath "smokes" in the winter. Our exhaling breath also contains carbon dioxide. In fact, it is 4½ per cent carbon dioxide.

This carbon dioxide is *not* poisonous in moderate amounts. The trouble with it is simply that it is "used-up" air. If too much of it accumulates, taking the place of oxygen, we would suffocate.

How does the body absorb the oxygen we breathe? The lungs, you see, are like two gigantic sponges. The air-tube that enters each lung breaks up into smaller and smaller branches until finally the air is led into microscopic pockets. These are called the ALVEOLI. These alveoli are spread all through the lungs. There are millions of them.

The alveoli have very thin walls. On the other side of the walls are tiny blood-vessels (known as CAPILLARIES) which also have thin walls. These walls are so thin that the air in the alveoli can leak right through into the blood inside the capillaries. In the blood are the red cells, billions of them. Each red cell is packed with the protein, hemoglobin. Every molecule of hemoglobin has the ability to combine *loosely* with four molecules of oxygen.

Blood flows to the lungs, therefore, and returns to the

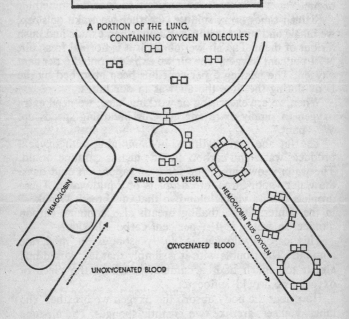

HOW THE BODY GETS OXYGEN

A PORTION OF THE LUNG, CONTAINING OXYGEN MOLECULES

SMALL BLOOD VESSEL

HEMOGLOBIN

HEMOGLOBIN PLUS OXYGEN

OXYGENATED BLOOD

UNOXYGENATED BLOOD

heart with its red cells loaded with oxygen. This "oxygenated" blood is sent out through the arteries to all the cells of the body. The oxygen, being held loosely, is given up to the cells. The "deoxygenated" blood then flows back to the heart through the veins and is sent back to the lungs for more oxygen.

Oxygenated blood is bright red. Deoxygenated blood is dark bluish-red. The blood we bleed is always bright red, because even if we bleed deoxygenated blood, it becomes oxygenated as soon as it hits the air.

There is a gas called CARBON MONOXIDE which is a close relative of carbon dioxide. Its molecule has one carbon atom and one oxygen atom (carbon dioxide, you remember, has two oxygen atoms.) The carbon atom is dissatisfied with only one oxygen atom and looks for another. For that reason, carbon monoxide will combine with oxygen. That is, it will burn in air. Gas used for cooking frequently consists of a mixture of carbon monoxide and hydrogen.

Carbon monoxide is so anxious to include something else in the molecule that if it can't get oxygen it will take something else. For instance, it will grab on to hemoglobin and stick to it *tightly*. It takes only a little carbon monoxide in the air to tie up most of the hemoglobin molecules in this way. Hemoglobin which is loaded down with carbon monoxide is unable to carry oxygen, without which the body cannot live.

That is why carbon monoxide is poisonous. It is the carbon monoxide that makes gas leaks in the kitchen so dangerous. It is the carbon monoxide in automobile exhausts that makes it important never to run your car motor in a closed garage. Hemoglobin combined with carbon monoxide is cherry-red and one way of telling that a person has died of carbon monoxide poisoning is by the pink flush on his face.

THE BUCKET-BRIGADE TECHNIQUE

Exactly how are fats and carbohydrates burned *slowly?* We can't manage to burn paper slowly. Either it doesn't seem to be burning at all, or else, if we put a match to it,

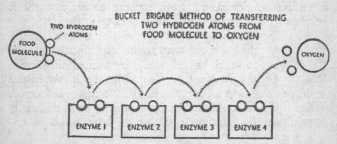

BUCKET BRIGADE METHOD OF TRANSFERRING TWO HYDROGEN ATOMS FROM FOOD MOLECULE TO OXYGEN

FOOD MOLECULE — TWO HYDROGEN ATOMS — OXYGEN

ENZYME 1 — ENZYME 2 — ENZYME 3 — ENZYME 4

it burns like fury. There is no way to slow the burning by putting a match to the paper very gently, or by using just a small flame. Once it starts, it goes.

The body uses a special system of taking a molecule apart, little by little. The molecule of paper, when it burns, just flies apart. All the carbons and hydrogens combine with oxygen helter-skelter. In the body, that would never do. Each molecule oxidizes in careful steps. Two hydrogens are removed at a time and combined with an oxygen atom to form a molecule of water. From what is left of the molecule, another two hydrogens are removed and so on.

As the hydrogens are removed, the molecule gets to consist mostly of just carbon and oxygen. When that happens a carbon and two neighboring oxygens break away and become a molecule of carbon dioxide.

How does the body manage to control this slow breakup of a molecule of fat or carbohydrate? We have the usual answer: enzymes.

In the previous chapter, we spoke of the large group of hydrolyzing enzymes. Now we have an equally large and equally important group known as the OXIDIZING ENZYMES. Almost all enzymes belong to either one group or the other.

When two hydrogens are removed from a molecule they are *not* instantly combined with oxygen. The body practically never works that simply. Instead a whole series of enzymes are involved. The first one takes the two hydrogen atoms from the molecule and passes it to a second. The second enzyme passes it to a third and so on. There may be as many as six different enzymes involved.

This whole process by which hydrogen is removed from a molecule and passed from enzyme to enzyme till it reaches oxygen is known as DEHYDROGENATION.

Why does the body go through all this trouble? In the first place, if the hydrogen were combined immediately with the oxygen, so much energy would be formed all at once that the body would not be able to handle it. When the hydrogen passes to an enzyme, just a little energy is formed. When it passes to the next enzyme, a little more makes its appearance. By the time it reaches the oxygen, all the energy has showed up, but in convenient packets that the body can handle nicely.

Secondly, when the hydrogen is passed from enzyme to enzyme, the process can be reversed. The hydrogen then passes backward to the original molecule. This reverse process is called REDUCTION and is often useful to the body.

All this can be compared to the bucket-brigade method of fighting a fire. Suppose there is a fire on the second floor of a building and the only water-supply available is a well in the back-yard. It would take too much time and energy to keep running downstairs with a single bucket, pumping it full of water and then running back. The house would burn down.

Instead, a series of men form a line from the well, into the house, and up the stairs. A series of buckets of water are pumped at the well and passed along the line from hand to hand. The process is much faster than the one man, one bucket scheme, and each man expends only a little energy at a time. Also, the buckets can move either way, up the stairs or (once the fire is out) down the stairs again. If the bucket brigade weren't there, that wouldn't be so. You could move the buckets one way by throwing them out the window, but you couldn't move them the other way unless you had a powerful arm and a good aim.

The oxidizing enzymes act like a bucket brigade, as you see, with hydrogen atoms taking the place of the water buckets.

The enzymes in the cell must be carefully arranged in order to make this system work. If every enzyme were just floating about loosely, the body could never be certain that the right enzyme would be in the right place at the right time. Enzyme 1 would take the hydrogen, then not find Enzyme 2 any place handy. It's as though the men in the bucket brigade were free to walk about as they pleased while they were fighting the fire. The house would burn down if that happened.

Well, in the cell, there are many small particles called MITOCHONDRIA. (See picture, page 42.) Each one of these mitochondria is a little bucket brigade all by itself. It contains all the necessary oxidizing enzymes lined up in the proper order and kept there. The hydrogen atoms just move along the line.

THE OXYGEN DEBT

Sometimes the body needs a lot of energy in a hurry. It needs so much, in fact, that it is impossible to breathe enough oxygen for the purpose. In such a case, the body has a trick for getting a little energy without oxygen. The process takes place mostly in the muscles. Naturally, when you are chopping wood, running upstairs or playing a snappy game of basketball, it is the muscles that need a lot of energy quickly.

This process of energy without oxygen makes use of the simple sugar, glucose. Glucose is *the* basic fuel of the body. Before the starch of the liver can be used, it must be broken down to glucose.

The muscles even have a special supply of starch all their own for quick conversion to glucose whenever necessary. Blood always contains a certain quantity of glucose and one of its most important duties is to carry this glucose to all parts of the body, for use as needed.

The glucose molecule, as we know, contains 6 carbon atoms, 12 hydrogens, and 6 oxygens. Well, it can be broken in half, so that it then becomes two molecules of LACTIC ACID. Each molecule of lactic acid has 3 carbon atoms, 6 hydrogens and 3 oxygens. (Lactic acid is the familiar substance that gives sour milk its taste and smell.)

When a glucose molecule is broken into lactic acid, no oxygen is used. However, the process of breaking liberates a little energy (about 5 per cent of what you could get by oxidizing glucose all the way to carbon dioxide and water). The breaking of glucose in half doesn't take place all in one step, but in a series of as many as a dozen steps. Each step is under the control of its own enzyme.

As you work, the glycogen supply of your muscles turns into glucose and the glucose turns into lactic acid. As the lactic acid piles up, you become more and more weary. Eventually, you are forced to stop and rest.

You must now get rid of the lactic acid somehow. To do that, you oxidize some of it to carbon dioxide and water. For that you need oxygen. The energy you get that way can be used to turn the rest of the lactic acid back into glycogen, and put you back where you started.

This is why, after exercise or work, you breathe rapidly while you are resting. You are supplying your muscles with the oxygen they need to get rid of their lactic acid. You are paying off your OXYGEN DEBT.

THE MIGHTY INITIALS—ATP

Sometimes energy can be useless. If a piece of coal burns in the open air, its energy is of no use to anybody. The heat and light spread out into the air and are gone. If it is burned in a furnace or under a pot, the heat can be put to use. If we put it under a boiler, we can turn water into steam and run a steam engine. If we use it to run an electric generator, we can produce useful electricity.

How does the body make use of the energy it gets from the oxidation of glucose? How does it keep it from being wasted? It can store energy in a form we have not yet mentioned: CHEMICAL ENERGY.

At this point, we must introduce a new chemical element: PHOSPHORUS. It is rarely seen as an element, though it occurs in compounds quite frequently. Its compounds are usually known as PHOSPHATES, and form an important part of soil fertilizers.

Half the phosphorus in the human body is found in the bones. In fact, one-sixth of the weight of the bones is phosphorus. The rest of the phosphorus is found in some of the most important protein molecules of the body. For instance, the gene proteins contain phosphorus. In addition, phosphorus is found in a compound called ADENOSINE TRIPHOSPHATE. Chemists have to talk about this substance so much that they have developed the habit of referring to it by by its initials. They call it ATP.

To understand the importance of ATP, we must think of the way in which atoms are held together to form a molecule. Actually, that is quite a hard thing to understand and modern theories involve a lot of advanced mathematics. We can simplify matters for ourselves, though, by imagining the atoms in a molecule to be little balls held together by rubber bands.

Most of these rubber bands are quite slack. If you were to cut them suddenly, nothing much would happen except that the molecule would fall into two pieces. A few of the rubber bands, however, are stretched tightly. Sometimes two particular atoms within a molecule can't get very close together for some reason. You can see that in that case, the rubber band would have to be stretched in order to reach from one atom to the next.

When such a stretched rubber band is cut, not only does

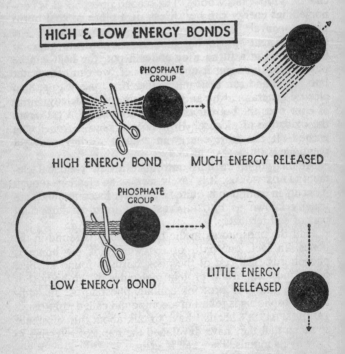

HIGH & LOW ENERGY BONDS

PHOSPHATE GROUP

HIGH ENERGY BOND   MUCH ENERGY RELEASED

PHOSPHATE GROUP

LOW ENERGY BOND   LITTLE ENERGY RELEASED

the molecule fall in two, but the force of the release causes the two pieces to spring apart violently. If you pull a rubber band taut between your two hands and have someone cut it, you will see that for yourself.

When two atoms in a molecule are combined so that the connection between them is like a slack rubber band, they are said to be held together by a "low-energy bond." When the connection acts like a stretched rubber band, it is called a "high-energy bond." Molecules that have such high-energy bonds possess considerable chemical energy.

The most important high-energy bond in living beings (plants, animals, and germs, as well as humans) involves the phosphorus atom. In ATP, there are three phosphorus atoms put together in a line at the end of the molecule. (Remember, the full name is adenosine *tri*phosphate.) The first phosphorus atom is connected to the rest of the molecule by a low-energy bond. The second phosphorus is connected to the first by a high-energy bond. The third phosphorus is connected to the second also by a high-energy bond. Although the molecule thus has two HIGH-ENERGY PHOSPHATE BONDS (as they are called), the body makes use of only one.

When the bond between the second and third phosphorus atoms is cut by hydrolysis, a considerable quantity of energy is set free for us by the body. The ATP molecules can be stored and hydrolyzed whenever needed. The little packets of energy each one releases are just the right size for use. The muscles, in fact, contain a special enzyme called ADENOSINE TRIPHOSPHATASE, whose only job is to hydrolyze ATP whenever energy needs to be transferred.

When the third phosphorus is cut off the ATP molecule, what is left is a very similar molecule, but one with only *two* phosphorus atoms. It is therefore called ADENOSINE DIPHOSPHATE, which is abbreviated to ADP.

Whenever glucose is oxidized, the energy is used to stick phosphorus atoms back on the ADP molecules in order to form ATP. The body cannot work at all except by using ATP molecules. ATP is the spark-plug of the body.

Another way of looking at it is to consider ATP as the "small change" of life-energy. The glucose molecule can be compared to a hundred-dollar bill. It is good money and can be used to buy many things. Still, you can't go on the subway with one, since nobody would have change for it. The ATP molecules are one-dollar bills. A hundred one-dollar bills are no more money than a single hundred-dollar bill. However, they are far more usable in ordinary life, because the problem of change isn't so important.

In fact, if you had a hundred-dollar bill you wished to use, your first act would be to go to the bank to change it. In the same way, the body changes its glucose energy into the handier ATP energy.

SUMMARY

The body continually uses up energy which it obtains by combining the carbon and hydrogen of the food we eat with the oxygen we breathe. This process is called oxidation and is controlled by a variety of special enzymes in the body. Oxidation takes place when hydrogen atoms from a compound in food are passed along a chain of different enzymes till it finally reaches oxygen. This is a sort of "bucket-brigade" technique. In the absence of oxygen, some energy can be obtained by breaking sugar molecules into two pieces. The energy obtained in all these processes is used to form "high-energy phosphate bonds" which are stored in the body as convenient chunks of energy for use whenever necessary.

6

The Friends and Foes of Enzymes

THAT NECESSARY TRACE OF METAL

IN THE PREVIOUS CHAPTER, we talked about oxidizing enzymes working in bucket-brigade fashion. We had enzyme molecules passing hydrogen atoms from one to another. Yet in Chapter 2 we said that all an enzyme did was to supply a proper surface for a chemical reaction. Why should we suddenly speak of enzymes as actually taking a personal part in a reaction? It's as though the chair we were sitting on suddenly leaned over and tied our shoelaces for us.

In order to explain this, we must begin by saying that *enzymes can't always do the job alone*. They have helpers and it is these helpers that can often take part in the reaction.

Some enzymes, for instance, require the presence of small quantities of certain metals. Atoms of these metals circulate in the blood and exist in the cells. Metal atoms which help enzymes in their work are called ACTIVATORS.

An example of such an activator is MAGNESIUM. Magnesium is a light metal, even lighter than aluminum. Its lightness makes it useful in airplane construction. Magnesium also has the property of burning in air with an intensely white light. Years back, photographers obtained a bright light for indoor snapshots by setting fire to "flash powder" consisting of powdered magnesium. Nowadays, we find it safer to use special "flash bulbs" containing fine curls of magnesium wire that can be set off electrically. The magnesium in the body, of course, exists in compounds which are not at all like the metal itself.

Other metals are also associated with enzymes in one way or another. These include the very common metals IRON and COPPER, which you all see every day of your lives. Three less common metals which act as activators are ZINC, MANGANESE and MOLYBDENUM. You may think you have never seen these, but if you tear the paper off an ordinary flashlight battery, you will find a gray-white metal on the outside which is zinc. Manganese and molybdenum are used in making tough steels, and are therefore very useful to industry and national defense.

The body uses very small amounts of these metals (which it gets out of the food we eat), but just the same they are necessary to life. The complete absence of any one of them would stop one or more vital enzyme reactions in the body and that would mean death.

Enzymes are often content to have their particular activators floating about freely in the cell. They call upon them only when necessary. In some cases, however, the function of an enzyme is so vital that there must be no risk of delay at all. The enzyme can't take the chance of having to wait at a crucial moment for an activator to float by. What it does, then, is hook its activator firmly to itself. The activator becomes an actual part of the enzyme.

This is particularly true when iron is the activator. Some of the most important oxidizing enzymes have iron activators built into their molecules. This is done in the following manner. The enzyme has, as part of its molecule, a compound known as HEME. Unlike the rest of the enzyme, heme is not made up of amino-acids. Heme is attached to the enzyme molecule like a postage stamp on an envelope and is shaped something like a flat square with a hole in the middle. In that hole an iron atom is neatly fitted.

Each of the two parts of the enzyme molecule has a function. It is the iron in the oxidizing enzyme that actually does the work of oxidation. But iron by itself couldn't do the job properly. The protein supplies the surface on which the reaction can proceed quickly.

If you want to bring all this down to familiar terms, consider ice skating. A man who wishes to move quickly and gracefully across the ice needs ice skates. To use the ice skates properly, he must clamp them firmly to the proper kind of surface. He uses the sole of his shoe for that.

Now we can consider the shoe as the enzyme and the ice skate as the activator. Each is useless without the other. In most cases, the ice skates are independent objects that can be adjusted to the shoe whenever needed. That is the system used by enzymes with free-floating activators. However, there also exist special shoes that have ice skates permanently attached. The enzymes that use iron as activator (or, as they are properly called, the HEME ENZYMES) are like the latter. The shoe is the enzyme itself; the permanently attached skate is heme; the cutting edge of the skate is the iron atom.

WHY A WHIFF OF CYANIDE KILLS

Now the iron is the key portion of any heme enzyme. Think of the ice skates and you'll understand. Your shoes may be run down at the heel and loose at the seams, yet you can still clamp on ice skates and go skating. If the cutting edge of the skates has been dulled, chipped and damaged, however, you cannot skate properly even though you have brand-new shoes of the finest quality.

There are certain substances that can ruin the workings of the iron atom and, consequently, prevent the heme enzyme from doing its job. One such substance is CYANIDE, which consists of a carbon and nitrogen atom hooked together. Whether cyanide is swallowed in the form of little pellets of potassium cyanide or breathed in the form of hydrogen cyanide gas, it can kill a man in a few minutes.

Cyanide attaches to the iron of the oxidizing enzyme and makes it useless, just as carbon monoxide makes hemoglobin useless. (Hemoglobin molecules, by the way, contain iron also. You might guess that from the name, in fact. Each molecule of hemoglobin possesses four heme groups complete with iron. Hemoglobin won't work without iron. That's why a shortage of iron in the food you eat results in a poor supply of hemoglobin in the body. Such a condition is known as ANEMIA.)

Cyanide acts much faster than carbon monoxide does. You see, there is a great deal of hemoglobin in the body and it takes a while for the carbon monoxide to load up enough of it to do us harm. The heme enzymes, on the other hand, are present in very small quantities. Just a few whiffs of cyanide are enough to ruin most of them. When that happens, the bucket-brigade which oxidizes the body's fuels stops dead. In a few minutes, a person's cells die of asphyxiation just as surely as though someone had seized that person's neck and choked him to death.

In fact, cyanide works so rapidly that it is used in the "gas chambers" of some of our western states to execute criminals who have been condemned to death.

A chemical such as cyanide, which prevents an enzyme from doing its work, is a POISON. Enzyme chemists prefer the longer name, INHIBITOR, and they call the process INHIBITION.

There is one important kind of inhibition that depends upon the fact that enzymes aren't perfect. Let us consider, for instance, an enzyme which needs magnesium as an activator. Although the enzyme is surrounded inside the cell by all sorts of atoms, it can tell magnesium from the rest without any trouble. At least that can be done under ordinary conditions.

However, there is a certain rather uncommon metal called BERYLLIUM, whose atoms are very similar to those of magnesium. In this case, the enzyme can't tell the difference and if beryllium is present in the cell, it gets picked up along with the magnesium. The trouble is that the beryllium is useless to the enzyme. It can't do its job with beryllium stuck to it. Worse than that, being filled up with beryllium, it can't even reach out to pick up magnesium, which it *can* use. It's just a ruined enzyme.

It's as though you wanted to open your front door on a dark night and had a key ring full of different keys. You can use only one for your front door and you can prob-

WHY CYANIDE IS A POISON

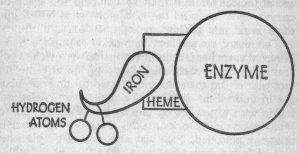

ENZYME

IRON

HEME

HYDROGEN
ATOMS

HEME-ENZYME AT WORK

HYDROGEN ATOMS UNABLE
TO FIND A PLACE

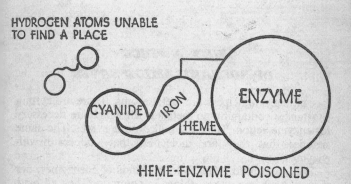

CYANIDE

IRON

ENZYME

HEME

HEME-ENZYME POISONED

ably tell which one that is just by the feel. But suppose you happen to have a key that very closely resembles the right one. You could easily make a mistake and put the wrong key into the lock. Then you would find it wouldn't turn. It might even break off in the lock when you tried to turn it so that the lock would become completely useless to you.

When two different metal atoms (or two of any kind of substance, for that matter) compete for space on an enzyme molecule, and the wrong one wins out, it inhibits the enzyme action. This is called COMPETITIVE INHIBITION.

The reason we use beryllium as an example is that only in recent years has it become recognized as a violent poison. It is a rare metal and ordinarily people never came in contact with it. But then factories started turning out tubes for use in fluorescent lighting. The tubes were coated on the inside with a powder containing beryllium. It was found that workers in factories making the tubes were sometimes seriously poisoned. That was also true of people who accidentally cut themselves on broken fluorescent tubes. As a result, we are now learning to take precautions against this new poison.

WHY A PINCH
OF SULFANILAMIDE SAVES

Not all enzyme helpers are metal atoms. There are certain substances containing no metals at all which are necessary for enzyme action. These are called COENZYMES. The name signifies that they are molecules that *co*operate with enzymes.

There are quite a few different kinds of coenzymes, but they all have one or two things in common. They all have molecules of intermediate size. That is, the molecules are larger than those of sugar but smaller than those of proteins. All the coenzymes seem to contain phosphate groups in their molecules. Usually these are so arranged that a high-energy phosphate group is present.

The best-known coenzymes are those belonging to the

bucket-brigade system we discussed in the previous chapter. The first enzymes in that series are helped by coenzymes which are most often called simply COENZYME I and COENZYME II. Coenzymes I and II resemble one another closely. Both of them have in their molecules a certain combination of atoms known as the pyridine ring. For this reason, enzymes that make use of these coenzymes are sometimes called PYRIDINOENZYMES.

The next enzymes in line make use of coenzymes which belong to a class of compound that chemists call FLAVINS. The enzymes that use them are therefore called FLAVOENZYMES. (The reason that these names are being mentioned now is that these coenzymes have an important connection with certain vitamins. We will begin the discussion of vitamins in the next chapter.)

The last enzymes in the bucket brigade are the heme enzymes which we discussed earlier in the chapter.

Now just as an enzyme can pick up the wrong metal and get itself inhibited, it can also pick up a slightly twisted coenzyme with the same bad results.

For instance, there is a certain coenzyme called FOLIC ACID, which is necessary to all forms of life, from mankind to bacteria. The middle part of this coenzyme has a shape something like that of a molecule of SULFANILAMIDE, which is a substance that chemists have known about for years.

In the middle 1930's, doctors discovered that if they gave sulfanilamide to patients with certain infections or diseases, the patients were cured. Sulfanilamide became famous immediately and most people are very familiar with its name now even though it does have five syllables.

The reason why sulfanilamide works is that some germs cannot tell the difference between sulfanilamide and the correct "middle part" of folic acid. When they want to build up folic acid for their own use, they will grab the sulfanilamide molecules as building material, if these molecules are handy. But once they build up folic acid with sulfanilamide as its middle part, they find they cannot use it. This throws some of their enzyme reactions completely out of kilter and they die.

Pneumonia germs are particularly likely to grab sulfanilamide and die. That is why sulfanilamide is given to patients with pneumonia. The doctor tries to give the germs every chance to make their little mistake. After doctors started using sulfanilamide, pneumonia lost most of its dangers. Many people are now cured who would certainly have died twenty years ago.

Doctors have used a variety of molecules of the sulfanilamide family. Each different one works in different ways and against different germs. All together, they are known as the SULFA DRUGS.

You may wonder why sulfanilamide doesn't kill human cells as well as bacteria, since folic acid is necessary to *all* life. However, although we do need folic acid for our bodies, we do not make it ourselves but get it ready-made in our diet. We have no chance, therefore, of making a sulfanilamide mistake. Still, if we use enough sulfanilamide, we *will* poison the patient. Fortunately, the pneumonia germ is more sensitive to sulfanilamide than our own cells are, so we can pick a quantity of the chemical which will be enough to kill the germ and not enough to kill us. All drugs become poisonous when too much is used. That is why you should never take medicine on your own, but always try to get a doctor's advice as to how *much* to use and how *often* to take it.

Sulfanilamide won't kill all germs. That is why doctors and chemists are always looking for new and better drugs. A germ that is immune to one drug might be duckmeat for another. Unfortunately, there are still many kinds of germs that can't be killed by any drug we know (without killing the patient, too). Luckily, the body has its own methods for fighting germs which often work when medicine can't.

In fact, there are cases where drugs can make bacteria more dangerous. Sometimes when a dose of drug is too small, there is only enough to kill the weaker bacteria. The stronger ones survive and start a new line consisting entirely of strong bacteria like themselves. After that, even the proper dose will no longer work and a stronger dose still may be dangerous for the patient. That is another reason for consulting a doctor when you are ill instead of trying to take medicine by yourself. Not only can you poison yourself by taking too much of a drug, but you could develop strong and dangerous germs by taking too little.

In the last ten years, a whole new group of substances for fighting infection has been discovered. These drugs are obtained from microscopic plants known as molds. The best known of these is PENICILLIN, which was isolated from ordinary bread mold.

Hundreds of such bacteria-killing drugs have been isolated. The newspapers call them "wonder drugs" or "miracle drugs," but they are more properly termed ANTIBIOTICS, from Greek words meaning "against life." Most of them, unfortunately, are not useful because they are against our life as well as against bacterial life. The trick is to find antibiotics which will act against bacteria and still be relatively harmless to us. There are only a few of these. In addition to penicillin, AUREOMYCIN and TERRAMYCIN have proven to be among the useful lifesavers.

Two others, streptomycin and chloramphenicol, are sometimes useful, but they are pretty near the borderline as far as harmfulness is concerned. Even doctors have to be very careful in prescribing them.

It is not yet known exactly how the various antibiotics work. However, just about everyone agrees that they must inhibit some key enzyme or enzymes within the cell.

Antibiotics are not effective against virus diseases such as measles, mumps, chicken pox or the common cold. Bacteria have enzymes of their own which can be attacked separately. Viruses, however, live inside the body's cells and use the cells' enzymes for their own purposes. To stop the enzymes which serve viruses, you must therefore stop the cells' own enzymes. That, obviously, would kill the patient.

The body frequently manufactures complex proteins that are particularly designed to combine with certain viruses and destroy them. These anti-virus proteins that the body makes are called ANTIBODIES. In some cases, the antibodies last you all your life and once you have a disease like measles or the mumps and recover, you rarely get it again. Other antibodies exist only temporarily and that is why you can catch other virus diseases like colds and the flu over and over again.

In the case of vaccination, doctors try to help the body make antibodies against a serious disease known as smallpox. This treatment is so effective that in the two hundred years vaccination has been used, smallpox has nearly been wiped out in many countries. Sometimes doctors try to supply antibodies ready-made. This is what they are doing when they use GAMMA GLOBULIN to prevent measles.

SUMMARY

Enzymes sometimes work along with metal atoms, which are called activators. For instance, iron is necessary in the working of some of the oxidizing enzymes. The iron atoms cannot work properly when a substance such as cyanide is present. Since enzymes are present in only small quantities, it takes very little cyanide to stop the working of enough oxidizing enzyme to kill a human being. That is why cyanide is poisonous. Other enzyme-helpers are not metals, but are fairly large phosphorus-containing compounds. These are called coenzymes and are related to some of the vitamins. The sulfa drugs interfere with the workings of some of the coenzymes in bacteria, and kill the bacteria in that fashion. The antibiotics probably act in the same way.

7

The B Vitamins

THE BODY'S
CHEMICAL TALENTS FALL SHORT

IN GENERAL, LIVING ORGANISMS manufacture for themselves those chemicals they need to carry on the body's work. Plants, in particular, make everything they need out of very simple substances. They start with nothing more than water, carbon dioxide, minerals from the soil and the energy of sunlight. Out of this, they put atoms together in patterns of infinite variety. They make starch, proteins, cellulose and uncounted other types of molecules.

Animals are less versatile. To make their own proteins and tissue substances in general, they usually start with glucose and amino-acids. To get these compounds, animals must eat plants. Or else they must eat animals that have eaten plants. No matter how animals seem to live on one another, in the long run, every animal on earth, including man, lives on plants.

Well, then, suppose the human body is supplied with glucose and amino-acids, can it make everything it needs without exception? Strangely enough, the answer to that is "No." And there are reasons for that "No."

In the first place, the body needs different amounts of the different chemicals that make it up. For instance, it needs a great deal of hemoglobin so it maintains a complicated machinery to manufacture hemoglobin out of amino-acids. On the other hand, the body needs only a tiny bit of each of the various coenzymes we mentioned in the previous chapter. This poses a problem for the body.

Take Coenzymes I and II as examples. As we said in the previous chapter, they contain the combination of atoms known as the pyridine ring. This is a ring made up of 6 atoms: 5 carbon and 1 nitrogen. Now the pyridine ring occurs nowhere in the body except in those two coenzymes and in a third coenzyme which we haven't had occasion to mention. It would seem a shame for the body to have to set up chemical machinery (complete with special enzymes) just to manufacture a grouping it needs so little of. Yet that little quantity it *must* have.

A similar problem exists in any large industrial plant. An automobile factory, for instance, must manufacture large quantities of frames, wheels, motors, and so on. It concerns itself chiefly with metal-work. The manufacturer sets up machinery for the purpose. He would hate to tie up a lot of money, however, by investing in equipment that would make such items as paint or seat-covering. It is much simpler and more economical to buy such minor items, ready-made, from other factories that specialize in paint or textiles.

The body does precisely the same. It manufactures its own protein but it gets its pyridine rings ready-made out of the food it eats. It makes no provision for manufacturing it for itself. In general, the more specialized an animal is, the less it troubles to manufacture minor items for itself. It depends on its diet more and more.

This has advantages and disadvantages. The advantage is that the cellular machinery that is saved can be converted to other, more important uses. To understand that, consider the automobile factory again. If it doesn't have to set up machinery to weave cloth, it has that much more floor space to devote to manufacturing automobiles.

The disadvantage is that every once in a while a person's diet does not contain enough of these key atom combinations that the body cannot make for itself. When that happens, the body is in trouble.

These atom groupings, which the body cannot make for itself and which it must get from the food it eats, are called VITAMINS.

VITAMINS AND ENZYMES

We can get down to cases if we consider a disease known as PELLAGRA. People with pellagra suffer from red and swollen tongues. Parts of their skin, particularly on their hands and lower arms, turn first red, then brown and scaly. There are other symptoms, too, that are even more unpleasant.

Pellagra was once a medical mystery. For a while, most doctors thought it was a germ disease. The trouble was that they couldn't locate the germ. What's more, pellagra didn't seem to be contagious. Then other facts turned up.

In the first place, as far as the United States is concerned, the disease occurs mostly among people whose diet is very restricted. Those who, out of poverty, lived mostly on corn and meal with not much of anything else were particularly likely to get pellagra. In the years before our modern knowledge of vitamins was developed people who were in prison or in asylums of some sort, where the diet was poor and monotonous, were frequently victims of the disease.

On the other hand, families who owned a cow or who could afford to buy milk, were fairly immune to pellagra. In fact, if a pellagra victim was able to drink milk every day, the disease gradually left him.

Biochemists and doctors now know that the body suffers from pellagra when it lacks sufficient pyridine rings. The pyridinoenzymes which use Coenzymes I and II run at low pressure or scarcely at all. The oxidizing machinery of the body begins limping. The swollen tongue, the scaly skin and other symptoms are the visible evidence of the body's failing chemistry.

Some foods contain pyridine rings in the form of a compound known as NIACIN. Good sources of niacin are yeast, liver, meats, fish, eggs, whole wheat, brown rice, and peanuts. Corn, on the other hand, has little or no niacin. Now this doesn't mean that corn is bad to eat. Corn is a very delicious food and deserves to be popular. However, it will not support healthy growth and development all by itself.

Niacin is our first example of a vitamin. The body needs only one two-thousandth of an ounce each day. This certainly isn't much, but it is vital. And, as you see, it is possible that you won't get even that tiny amount if your diet is an unwise one.

A disease, such as pellagra, that is due to a vitamin

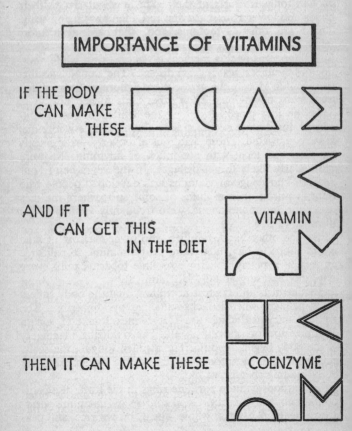

deficiency is called an AVITAMINOSIS. Pellagra is the most frequently occurring avitaminosis in the United States.

In other parts of the world, where dietary problems are greater than in the United States, avitaminoses occur more often. In oriental countries, among people whose diet consists largely of polished white rice, a disease known as BERI-BERI occurs. This has many serious symptoms, including nervous disorders and heart trouble. People who eat unpolished brown rice do not get it.

In the outer coatings of the rice kernel, there exists a chemical known as THIAMINE. This cannot be made by the body, but it is a necessary part of the coenzyme that services one of the important enzymes of the body.

Unfortunately the outer coatings of rice also contain oil, so that unpolished rice is likely to become rancid when stored. The East Asian peasants, in order to avoid rancidity, get rid of the outer coat by "polishing" the rice. If their diet consists mostly of this polished rice, they suffer from an avitaminosis, the one that we have already referred to as beri-beri.

Wheat, like rice, contains thiamine (and other vitamins) in the outer portions of the grain. That is why whole wheat bread contains more vitamins than does white bread which is made out of grain with the outer layers removed. Manufacturers of white bread know this and nowadays they *add* thiamine and other vitamins to their products.

Still a third vitamin is RIBOFLAVIN. It is needed for the coenzymes servicing the flavoenzymes we mentioned in the previous chapter. Milk is the best everyday source of riboflavin.

The human body's daily requirements for thiamine and riboflavin are even less than that for niacin. It can get along with only one twenty-thousandth of an ounce a day. (To put it another way, an ounce of niacin, if taken bit by bit every day, would last the body for six years. Similarly an ounce of riboflavin or of thiamine would last it for sixty years.)

All three of these vitamins belong to the group known as the B-COMPLEX VITAMINS. In fact, thiamine is often called Vitamin B_1, and riboflavin is often called Vitamin B_2.

The system of naming vitamins by letters and numbers arose quite by accident. When it was first discovered that certain chemicals in the diet were necessary for human health in very small quantities (*i.e.* "traces"), no one knew what they were. It took years of careful study to isolate them and work out their chemical structure. Meanwhile, they were just called "vitamins" from a Latin word meaning "life."

At first, only two kinds of vitamins were thought to be present in food. One was soluble in fat but insoluble in water. This was called Vitamin A. The other was soluble in water but insoluble in fat. That was called Vitamin B. It turned out, however, as knowledge increased, that what was called Vitamin B was not a simple substance. Actually, it proved to be a mixture of a number of vitamins with similar chemical properties. All contained nitrogen. All were soluble in water. All formed parts of coenzymes.

For a while, they named the members of this B-complex by attaching numbers: B_1, B_2 and so on. Once the chemical structure of each vitamin was determined, however, a name was attached. In the case of the B-vitamins, at least, the letter-number system is going quite out of fashion. This is true even among the general public. Manufacturers are stressing the vitamin content of their food products so much these days that Americans are becoming familiar with such words as thiamine, riboflavin, and niacin.

VITAMINS AND DIET

The three vitamins we have just talked about are the "major" B-vitamins. They are the only members of the B-complex that are likely to present a dietary problem. Avitaminoses involving the remaining half dozen or so B-vitamins are rarely reported.

In order to study the minor members of the B-complex, it is necessary to work with artificial diets. These diets contain purified amino-acids, refined oils and sugars and so on. They contain no vitamins or minerals except those which the experimenter deliberately adds.

Obviously, these diets can't be used on human beings very well. In the first place, dietary experiments may continue for months and artificial diets are expensive. Secondly, humans would revolt at such tasteless, uninspired and monotonous food and would be too likely to sneak down to the corner for a hamburger or a milk-shake. That would ruin the experiment.

So instead the biochemist makes use of animals. Usually, the animal he chooses is the rat. The rat is easy to handle. It breeds quickly so that there is always a good supply. It is small so that it doesn't eat too much. Furthermore, it is one of the few animals which, like man, will eat almost anything. This enables experimenters to check on the vitamin contents of ordinary human diets.

Of course, no animal resembles man perfectly. For instance, rats develop pellagra only with great difficulty so that they are useless for checking on niacin contents of food. In the case of that vitamin, dogs make better laboratory animals. Germ growth (in test-tubes, of course) has also been used as a way of studying vitamins.

At least five B-vitamins have been identified as a result of experiments with laboratory animals. These are PANTOTHENIC ACID, PYRIDOXAL, BIOTIN, FOLIC ACID, and COBALAMIN.

The American public is very vitamin-conscious. Millions of dollars are spent in advertising vitamin-enriched food and vitamin pills. As a result, the average American is likely to think the danger of avitaminosis to be greater than it really is.

Actually, the B-vitamins are found in all cells and hence in all natural food (though not always in sufficient quantity). Naturally, if we throw away portions of natural foods, such as the outer layers of rice or wheat kernels, we may be throwing away vitamins.

Just the same, anyone who eats a varied diet in reasonable quantities is unlikely to have trouble with his B-vitamins. In general, the best sources for the B-vitamins are those tissues which are engaged in considerable chemical activity. These tissues require a more than average amount of coenzymes, you see, so they have stored up the vitamins that form part of the coenzymes. Meat, liver, and yeast are examples of such chemically active cells or tissues.

Other good sources include the food supplies intended for the use of organisms before birth or shortly after. For instance, eggs contain all the food elements necessary for a three-week growth and development. In that period of time, the spot of life inside the egg must grow from a single cell to a young chick. Naturally, eggs must have a good supply of B-vitamins if the chemical machinery is to do all that work in three weeks. (It is the yolk of the egg that has it, by the way, not the white.)

For the same reason, milk and seeds are good sources of B-vitamins.

The method of preparing food can affect its vitamin content. The two chief vitamin hazards in preparing food are, one, heating, and two, soaking in water. The B-vitamins in general are not much affected by heat, so that part is all right. However, since they are water-soluble, they are liable to be soaked out of the food into the water used in cooking. Broiled meat has therefore retained more vitamins than has boiled meat. It is also for this reason that soups and drippings can be good sources of B-vitamins.

VITAMINS AND BACTERIA

No matter how we scrub and bathe ourselves, we are loaded with germs all the time. We carry germs not merely on the surface of our bodies, but also in our mouths and, most of all, in our intestines.

By and large, these germs do us no harm. We and the germs have learned to live together. The germs in our intestines make use of the fragments of food reaching them that have not yet been digested and absorbed. In turn, they refrain from invading more delicate parts of the body or from producing poisonous substances to make us ill.

The existence of intestinal bacteria explains a recent finding which may have important effects on the world's food supply. When small quantities of antibiotics such as aureomycin are added to the feed of pigs, lambs, or chicks, they gain weight at a more rapid rate than usual. This probably results from the effect of the antibiotic in suppressing the growth of bacteria. If there is less bacteria in the intestines, there is less food absorbed by them. More food remains available for the animal itself. Therefore it grows faster.

Does this mean that it would be a good idea to use the new wonder drugs to kill the germs in our intestines? No. Those germs actually do us good.

After all, bacterial cells succeed in living and to do so they must have a complete set of enzymes and coenzymes. The more common types of bacteria in our intestines are more versatile, chemically, than we are. They can manufacture some of the B-vitamins for themselves out of simpler chemicals, a thing we cannot do. The bacteria manufacture so much that some of it leaks out of their cells into the intestines. The human body can then absorb it. In fact, it is quite possible that the human body can get all the minor B-vitamins it needs as a result of the activity of our bacteria, even if there were absolutely none in the diet.

This accounts for the strange fact that there are vitamins in body wastes. In fact, in the case of some vitamins, there is more in the wastes than in the original food. In experimenting with rats, it is sometimes necessary to keep them on wire nets a few inches above the bottom of the cage. In this way their droppings fall through the holes of the nets and can't be reached. Otherwise, by eating their own droppings rats might obtain some vitamins and ruin the experiment.

Even in the case of the major B-vitamin, niacin, bacterial activity is important. Bacteria can form it from an amino-acid called TRYPTOPHANE. In this way, we get more niacin than we would expect, if we consider only the quantity in our diet. In fact, the reason why pellagra results from a diet consisting chiefly of corn may not be entirely due to the fact that corn is poor in niacin. Unfortunately, the proteins of corn happen to be poor in tryptophane, too, so that even bacteria can't help us there.

In the next chapter, we will come across a case where intestinal bacteria can actually prove to be necessary to life.

THE NEWEST VITAMIN

A few pages back we mentioned cobalamin as one of the minor B-vitamins. You may not have recognized the name. It is better known by the old-fashioned name of Vitamin B_{12}. It is the latest important vitamin to be discovered.

To understand the importance of B_{12}, we must first consider a disease known as PERNICIOUS ANEMIA. Anemia is a general name for any disease in which there is less than a normal amount of hemoglobin in the blood. Most frequently, it is caused by lack of sufficient iron in the diet. (Iron, you will remember, is a vital part of hemoglobin.) Simple anemia can be cured by taking pills made up of chemicals containing iron atoms, or by eating more meat and eggs.

There are more serious anemias of many types, however. After all, the process of manufacturing hemoglobin in the body is a complicated one with many steps. Interference at any step in the process will result in anemia. Pernicious anemia is one of the most serious forms. Up to about thirty years ago, it always resulted in death sooner or later.

It was then found that certain types of liver preparations held this disease in check. Still later, after World War II, it was found that the disease was due to a lack of a vitamin. This vitamin was called first B_{12}, then cobalamin. The trouble with pernicious anemia patients was not that the vitamin wasn't in the diet. In fact, bacteria manufacture it, so the diet problem isn't important. For some reason, though, pernicious anemia patients cannot absorb cobalamin into their bodies. The injection of cobalamin ends their troubles, however.

There are several very interesting points about cobalamin. In the first place, it is needed by the body in very small quantities indeed. One fifteen-millionth of an ounce per day is enough to keep a pernicious anemia patient healthy. Normal people can get along with half that. In other words, an ounce of cobalamin would last the entire population of Cleveland, Ohio, for a whole month.

Then, too, cobalamin has a more complicated molecule than any other vitamin. It was only in 1956 that the formula was worked out. The experimental data that had to be handled before the formula could be decided upon was so complicated that a large "mechanical brain" had to be used to interpret them. It turns out that cobalamin has a molecule that resembles heme somewhat, but is considerably more complicated.

Finally, cobalamin is unique in that each molecule contains an atom of the metal, COBALT. In fact, that's where it gets its name from. The cobalt atom fits into the center of the cobalamin molecule just as the iron atom fits into the center of the heme molecule.

Cobalt is a metal which is much like iron but considerably rarer. It occurs along with iron in certain ores. Miners in medieval Germany who came across it were much disturbed. It acted something like iron, but not quite, and it got in the way of their smelting operations. Their theory was that evil spirits of the Earth had somehow enchanted the iron and made it useless. Such evil spirits were called "kobolds" and so the new metal got its name.

SUMMARY

The body cannot manufacture all the compounds it needs out of the simple substances it gets by digesting fats, carbohydrates and proteins. The additional compounds it needs must be gotten ready-made in the diet, and they are known as vitamins. They are only needed in small quantities. The B-vitamins are a group of vitamins which are necessary because the body has to make certain necessary coenzymes out of them. If they are absent in the diet, the body cannot make the coenzymes and several important enzyme-controlled reactions slow up. The result is illness and even death. To make sure we get enough vitamins, our diet should be varied and include milk, eggs, meat, fruit and fresh vegetables. Some B-vitamins are manufactured in sufficient quantity by bacteria in our intestines so that we don't have to worry about their presence in our diets. Vitamin B_{12} is the most recently discovered vitamin. It contains the metal, cobalt, and is useful in preventing a serious form of anemia.

8

Other Vitamins

VITAMIN A HELPS US SEE

THERE ARE QUITE A FEW VITAMINS that do not belong to the B-complex. None of them contain nitrogen. Most of them are insoluble in water. In general, we know less about their actual functions in the body than we do in the case of the B-vitamins.

We can begin with Vitamin A. It comes first alphabetically and it was one of the first discovered. Back in 1913, it was found that some fats and oils stimulated rat growth, while others did not. At the time, there didn't seem to be any chemical difference involved. Afterwards, however, it was discovered that the growth-stimulating fats contained very small quantities of the fat-soluble Vitamin A. Despite this long history, the vitamin is still known almost exclusively by its letter. No chemical name has caught on.

Most people connect Vitamin A with carrots, and that's shrewd of them. Carrots are a good source of the vitamin, even though they contain practically none of it.

What carrots do contain is a substance known as CAROTENE. Carotene has a long molecule composed exclusively of carbon and hydrogen. There are a whole family of compounds related to carotene. One of the family resemblances among them is that they are colored. The colors vary from yellow through orange to red. Many foods owe their color to the carotene family. For instance, butter is yellow, tomatoes are red, and carrots are orange. The carotene family is responsible in each case and all three foods are good sources of Vitamin A.

The yellowish color of many fats and oils is due to their carotene content. Even the human being stores carotene in the fatty layers that lie just under the skin. If a considerable amount is stored, human skin can have a noticeable yellow tinge, which is the case among the people of eastern Asia.

What is the connection between carotene and Vitamin A? Well, the carotene molecule, as we have said, is a long one. It contains 40 carbons, mostly in single file. The human body can break that molecule into two equal pieces of 20 carbons each. It heals the point of breakage by putting one atom of oxygen and one of hydrogen on each piece. When it has finished this job, it has obtained two molecules of Vitamin A.

In other words, carotene, although not a vitamin itself, can be converted into one by the body. Carotene is therefore called a PROVITAMIN.

Vitamin A can be stored in the body to a greater extent than the B-vitamins. You may have been thinking of this. If vitamins are so necessary, you may have thought, why doesn't the body "stockpile" them when they happen to be plentiful in the diet? Later on, during lean seasons, the body can fall back on its supply. Actually, this isn't as easy as it sounds.

Let's consider a factory again. Suppose it decided to stockpile certain items for use during strikes or shortages. In the first place, it would have to find room. Then it would have to hire bookkeepers to keep track of exactly how much of each item was in stock. And the owners would have to invest money in the stored items.

The body has similar problems. It can't just toss vitamins into some unused corner. It has to get up special enzyme mechanisms to keep vitamins stored within certain cells that can manage to adjust their own chemistry to make room for them. It's rather complicated to do. In most cases the body seems to find it easier to just rely on the diet.

In the case of Vitamin A, however, some storage does take place. It is stored in the liver. For that reason, liver is a better source of Vitamin A than ordinary meat is.

Sea animals seem to store Vitamin A to a larger extent than do land animals. In fact, before the days of modern vitamin concentrates, children used to be given regular doses of cod-liver oil. As the name suggests, this was the oil which was squeezed out of codfish livers. It contained large quantities of Vitamin A. Even richer supplies could be drawn from the liver of halibut.

Fish aren't champion in this respect. The polar bear is. In fact, polar bears store such large quantities of Vitamin A that their livers are actually poisonous. Explorers who ate polar bear liver have been made ill and actual deaths have been rumored.

This may surprise you. How can vitamins hurt you when they are so necessary to life? Well, water is necessary, too, but you can still drown in it.

There *is* such a thing as too much vitamins. The B-vitamins are relatively non-poisonous. At least, you can eat many times what you need without bad effect. Of course, eating too much doesn't do you any good. Once your body has all the B-vitamins it can hold, it just pours the excess out in the body wastes. It's like adding water to a full glass. It just splashes down the drain.

Vitamin A is another affair. It can be stored in sufficient quantity to get in the way of the body's workings. Every once in a while, doctors get a little boy or girl as a patient. The mother has been told to give the child ten drops of vitamin solution every day. She imagines she'll have a really healthy baby if she gives him a teaspoon every day. The child gets sick. Sometimes the liver is damaged.

Perhaps you have heard that the British fliers ate carrots during the Battle of Britain to improve their eyesight for night-flying. This was because Vitamin A has much to do with the chemistry of sight. It has other functions as well, but the best-known effect of a deficiency of Vitamin A is "night-blindness." The patient cannot see in dim light. In severe cases, the membranes about the eye become thick and dry. This condition is called XEROPHTHALMIA. (The "x" is pronounced like a "z.")

The eye is much like a camera. Light enters through a transparent layer of tissue, called the cornea. Behind it is the iris (the colored part of the eye) which by expanding or contracting controls the quantity of light. The light passes through the pupil (which is black because it is a hole leading into the eye's interior). It then goes through an adjustable lens which works exactly the way a lens in a camera does. The lens focuses the light on a very delicate tissue in the back of the eye's interior, called the retina. The retina corresponds to the film of the camera.

The retina is composed of two kinds of cells: the rods and the cones. The cones are used for seeing in bright light. They also see in full color. The rods specialize in dim light and see in black and white.

The rods contain a chemical called RHODOPSIN. Because of its color, rhodopsin is frequently called "visual purple." Rhodopsin is made up of two parts: a protein, called opsin, and an auxiliary group, called retinene. Retinene is Vitamin A with two hydrogens removed.

Now when light hits rhodopsin, it splits up. The retinene comes loose. In the dark, the two parts join up again. It is that process which enables us to see in dim light. When the retinene is separated from the rhodopsin, some of it breaks down and is destroyed. It must be replaced from the Vitamin A stores in the body. When the body runs out of Vitamin A, the system breaks down. Night-blindness results.

VITAMIN D AND SUNLIGHT

Another group of compounds in the body that we have not yet mentioned are the STEROIDS. These are about half the size of the carotenoids, and have only 18 to 28 carbons in the molecule. Moreover, the carbons, instead of being lined up in single file, are arranged in four connected rings.

Different steroids vary from one another by slight differences in the arrangement of the atoms in their molecules. They differ also in oxygen content, which can vary from 1 to 5 atoms per molecule.

Many of the steroids are present in the body in only the smallest traces and yet are very important. Exactly how important is a question we are going to leave for the last chapter of the book. At this point, we will consider only one of the steroids, a compound called CHOLESTEROL.

Cholesterol is far from being a minor constituent of the

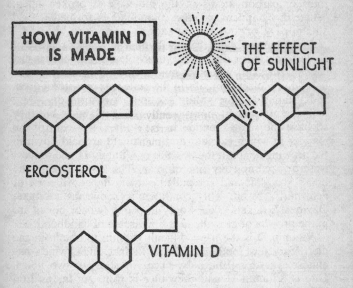

body. For instance, one-tenth of the solid substance of the brain is cholesterol.

Then again, liver secretion, or bile, is very rich in cholesterol. It is so rich, in fact, that sometimes it carries more cholesterol than it can really hold. The cholesterol settles out, therefore, and gradually collects into hard little masses. These are called "gallstones." Sometimes they block the duct that leads from the liver to the intestine, causing considerable pain.

Despite the fact that cholesterol occurs in such quantity, its function in the body is unknown. Just think! One-tenth of the brain solids are cholesterol and we don't know what it's doing there!

We're better informed about certain close relatives of cholesterol. One such is ERGOSTEROL, which has one carbon atom more than cholesterol. While cholesterol is found only in animal bodies, ergosterol is found only in yeasts, molds and so on. Now ergosterol has an important property which cholesterol does not have. It is a provitamin.

When ergosterol is exposed to sunlight, one of the four rings of carbon atoms in the molecule is broken open. When this happens, a new compound is formed, called Vitamin D_2, or ERGOCALCIFEROL.

In the human body, there is found a still closer relative of cholesterol. It is called 7-dehydrocholesterol. It occurs just beneath the skin. When the human body is exposed to sunlight, that molecule breaks in the same way ergosterol does, and Vitamin D_3, or CHOLECALCIFEROL, is formed.

There are several more varieties of D-vitamins, but they all have the same function in the body. They control the manner in which calcium and phosphate are laid down in the growing bone. In the absence of Vitamin D, the process is thrown completely out of gear. The result is that the bones stay soft and get pulled out of shape. All sorts of deformities result. This condition is known as RICKETS. Obviously, rickets occurs only in people whose bones are in the process of growth, so it is a disease of childhood.

Vitamin D is the most difficult vitamin to obtain in the diet. Very few foods contain it. In fact, milk, which has almost everything the baby needs, is short in two items. One is Vitamin D and the other is iron. As far as iron is concerned, the baby starts life with a six-month supply that the mother has contributed. By the time the six months are up, the child is probably eating at least some eggs or cereal, and these are good sources of iron. So that's all right.

Vitamin D is a more serious matter. For this reason, most brands of milk, at least in the United States, have had Vitamin D added to them. Furthermore, it is quite common to give very young children regular doses of vitamin concentrates containing Vitamin D.

Actually, though, before modern advances, it was the sun that saved children. Even though Vitamin D is rarely found in food, the body has no trouble in preparing the provitamin, 7-dehydrocholesterol. Human skin always has it. It is only necessary to stand in the sun, and the sunlight will convert it to Vitamin D_3.

That is why people call Vitamin D the "sunshine vitamin." It is not actually in the sunshine. Nothing material is in sunshine. It is merely that sunshine helps us make our own vitamin.

Articles of food, such as bread or milk, can be exposed to sunlight or to artificial lights that imitate sunlight. When this is done, provitamins in the food are changed to Vitamin D. The food is then said to be "irradiated."

There is a theory that the Vitamin D problem is what makes some people in northern climates have such pale skins. In northern climates, the sun is low in the sky and sunlight is weak. You need a colorless, semi-transparent skin so that the weak sunlight can penetrate and keep you from dying of rickets. Even so, rickets is found only too often among the young children of Europe, particularly those who are born at the beginning of winter.

In tropical climates, however, rickets is not such a problem. In fact, you need a dark complexion to keep from getting badly sunburned.

Vitamin D, like Vitamin A, can have bad effects on the body if taken in excess. Bone forms too readily, even in places where it shouldn't. In the joints, for instance. This results in stiffness and considerable pain.

THE MYSTERIOUS VITAMIN C

Man's most exclusive vitamin is ASCORBIC ACID. (A more old-fashioned name for it, but one which is still quite popular, is Vitamin C.) All plants and almost all animals can make their own Vitamin C. About the only exceptions are man, the apes, and the monkeys. There is only one other animal that cannot make Vitamin C. That is the guinea pig.

Ascorbic acid is very similar chemically to the simple sugars. It is the only important vitamin outside of the B-complex which is soluble in water.

Of all the vitamins, ascorbic acid is present in the body in the greatest quantity. It seems ironic, then, that of all the vitamins, it should be the least understood. We have a pretty good notion of how the B-vitamins work. We know at least some of the ways in which Vitamin A and Vitamin D fit into the body's chemical machinery.

About the workings of ascorbic acid, however, we know almost nothing. Of course, we know what happens when the body fails to obtain Vitamin C. Small blood vessels get weak and break easily, so that the patient bruises at almost the slightest blow. The gums, especially, bleed and become puffy. Wounds, in general, heal very slowly. This condition is called SCURVY.

Scurvy was perhaps the first vitamin disease to receive intelligent attention. The reason for this is that people frequently got scurvy in the old days of sea travel. Voyages by sailboat could take months and months. The diet would have to be restricted only to foods that could keep that long: dried beef, biscuits and so on. Unfortunately such foods contain little or no ascorbic acid.

Long before there was any knowledge of vitamins, an official of the British Navy discovered that if the sailors ate limes during voyages, scurvy did not strike. That is why British sailors are still called "limeys" and why there is a section of the London waterfront called Limehouse.

Ascorbic acid is found in many fresh fruits as well as in fresh leafy vegetables. The most used sources in the United States are certain fruit juices: orange juice, grapefruit juice and tomato juice. Drinking such juices has become such a national habit that scurvy is no longer a very likely disease. Because milk is somewhat short on ascorbic acid, orange juice is particularly important for babies.

Vitamin C is the one vitamin which is absent in animal foods. Since most make their own, they have no need to store it. They just make it as needed. The human body, unfortunately, has forgotten how to make it, but hasn't learned how to store it to make up for that.

Vitamin C is the most fragile of all the vitamins. It slowly disappears as food stands. This is why the best sources are *fresh* fruits and vegetables. Also, it is easily destroyed by heat, so that cooked vegetables have considerably less of it than do raw vegetables. Since it is water-soluble, it is also extracted by the water used in cooking.

Fortunately, in America the main source is the orange, which is neither cooked nor soaked in water. In many parts of Europe, however, the chief source of ascorbic acid is the potato. The potato doesn't have much Vitamin C in it, but it has a little. If enough potatoes are eaten, that little bit may stand between you and scurvy. In the case of potatoes, cooking lowers the margin of safety. (Incidentally, Vitamin C is usually found mostly in the outside layers of vegetables. Peeling potatoes with too heavy a hand, for instance, results in most of the vitamin being thrown away with the peel.)

Despite the fact that the human being doesn't store Vitamin C very well, it takes months of deprivation to get scurvy in a serious form. The reason for this is that the body can learn how to become economical with its vitamins.

We may take an analogy. If a man is earning $5,000 a year and has $5,000 in the bank, it seems logical to suppose that he can live on his savings for a whole year after losing his job. Actually (assuming there is no inflation), he can live longer than that. As soon as he is out of a job he begins economizing. He makes less do. He makes every penny count.

It's the same with the body. Somehow it manages to cut down when Vitamin C (or other vitamins) are slow in coming in. Eventually, it *is* in trouble, but the margin of safety is greater than some people might think. Vitamin-conscious mothers should remember that for their own peace of mind. Young mothers sometimes forget to give their babies their regular vitamin drops one day and are worried sick lest the children wake up deathly ill the next morning.

VITAMIN K AND BLOOD

Although several more vitamins exist, only one other, besides those we have already mentioned, has a definite use as far as the human body is concerned. We will end the roll with it.

To do so properly, we must bring up the subject of blood.

Now blood is a fascinating liquid with many remarkable properties. We have mentioned it before in connection with hemoglobin. Actually, that doesn't even scratch the surface. Whole books can be written about blood, and have been written.

One important property of blood is its ability to clot. You probably take that ability for granted. Countless times in the average person's life, scratches and scrapes have resulted from the ordinary accidents of everyday life. Blood may flow for a while, then it stops. The blood thickens in the wound's opening and seems to jellify. It forms a hard clot or "scab." After a while, the scab falls off and new skin is found underneath. The wound has healed.

Such a self-sealing device is obviously important if we're to stay alive. We can't have all our blood pouring out every time there's a break in the skin. On the other hand, you don't want blood clotting while it's still inside the blood vessels. In order to make the process as fool-proof as possible, to make blood clot only where and when necessary, the body has worked out a rather complicated procedure.

The exact details don't matter but a whole series of reactions are involved, including several enzymes, several ordinary proteins, and other chemical substances. New refinements are still being discovered.

The point is, though, that any hitch anywhere along the line can result in serious bleeding difficulties. Some people are born with imperfect clotting mechanism. Such a condition is known as HEMOPHILIA. The disease runs in families. People with hemophilia just keep on bleeding and even the smallest scratch might kill them eventually, if they do not receive expert care. In the last century, hemophilia has cropped up among members of the Spanish and Russian royal families. The disease therefore received considerable publicity. It still does even though neither royal family is royal any longer.

Now one of the steps in the clotting mechanism involves the compound we call Vitamin K. If Vitamin K is not present in the body, clotting cannot take place. In fact, Vitamin K received its "K" from the word "coagulation" (which is a fancy synonym for "clotting"). It was a German chemist who named it and in German, you see, the word is "koagulation."

Vitamin K is one of the vitamins which is most readily manufactured by the bacteria of the intestine. It does not have to be in the diet at all. Because of this, the only human beings who have to concern themselves with Vitamin K are *new-born babies*. The trouble is that babies are born without bacteria. It takes about three or four days for ordinary contact with the outside world to fill their tiny intestines with germs. Until then, they are in danger because they have no Vitamin K.

In other words, for the first three or four days of its life, a baby is a "bleeder." It has a mild and temporary hemophilia. And all because it has too few germs.

Modern American hospitals avoid this condition by injecting the mother with Vitamin K shortly before the baby is born. The Vitamin K leaks over from the mother's bloodstream into the baby's, and then the little fellow is safe.

Another interesting thing about Vitamin K is that chemists have created an unusual synthetic substitute for it. The molecule of the substitute is like that of the real thing except that it is simpler. Whereas Vitamin K has two rings of carbon atoms in its molecule with a long "tail" of more carbon atoms attached, the substitute has merely the rings without the tail. The interesting thing about it is that the substitute is over a hundred times as effective as the real thing. So here is one place where the chemist has outdone nature.

SUMMARY

In addition to the B-vitamins, there are several others of importance. Vitamin A is necessary to certain chemical reactions that proceed in the eye. Without it we cannot see properly in dim light. Vitamin D is the rarest vitamin and is found in very few foods. The best sources are certain fish oils. There are compounds in our skin, however, which can be changed to Vitamin D on exposure to the sun. Vitamin D is necessary for the proper formation of bones. Vitamin C is found in orange juice and in certain other fresh fruits and vegetables. It is necessary to keep the body's small blood vessels strong and well-formed. Vitamin K is necessary to the proper clotting of blood. It is formed by bacteria in the intestine. Since new-born babies don't have such bacteria it is now given to women who are about to have babies or to the babies themselves.

9

Protein Hormones

GLANDS AND SUPERVISORS

THE PICTURE THAT WE HAVE DRAWN of the human body so far in the book is incomplete. We have pictured billions of billions of cells. In each one of the cells, thousands upon thousands of chemical changes are occurring all the time. With all that activity going on, the result could easily be complete confusion.

Suppose we go back to our favorite comparison and think of the body as a factory. If a factory had a thousand workers, each one of whom did a different job, work would quickly become impossible without foremen and supervisors. After all, some departments in the factory might have a great deal to do at times when others were slow. A special job might come in which would make overtime necessary for some of the workers but not for others.

That is where executives come in. They distribute the work. They make sure every part of the factory is doing its share; no more and no less.

It is even more necessary to organize the work of the body's enzymes, since they are more numerous than the workers in any factory. For instance, during infancy and

111

childhood, a human being must grow. That means that the protein-forming enzymes must work particularly hard. Once a human is adult, he stops growing. The protein-forming enzymes slow down. Who or what tells them to?

When a boy is in his teens, parts of his body suddenly start growing faster than other parts. For instance, his larynx or "voice-box" begins to grow. As a result, his vocal cords grow longer and his voice grows deeper. Then hair starts growing on his cheeks, chin, and upper lip where none grew before. What causes this?

Suppose that you are suddenly frightened. Your heart begins to beat faster. Your breathing speeds up. Your muscles get ready to use up energy at a rate much higher than usual. You become less sensitive to pain. In short, your body is put on an emergency basis. You can run faster or fight harder than you could if you were not frightened. What makes the various parts of the body co-operate so neatly in this way?

The answer to all this are certain chemicals known as HORMONES. Hormones are produced by various organs called glands. We have mentioned glands in Chapter 4, but it is time now to be a little more detailed.

Glands are organs that produce fluids of one sort or another. Some glands possess little tubes, or DUCTS. The fluids produced by the glands pour through the ducts into some place where they are useful. For instance, the pancreas, which we have mentioned before, produces pancreatic juice which travels through a duct into the small intestine. There, the juice helps to digest food.

In the skin are millions of little glands that produce a watery fluid. The fluid travels through ducts to the surface of the skin. By drying on the skin, it cools us off in hot weather. If the air is so moist that the fluid cannot dry quickly, it collects in droplets on the skin. We then call it perspiration.

At the root of each hair on our body are small glands that produce an oily substance. This reaches the surface of the skin through ducts and helps keep the hair glossy, flexible, and water-repellent.

In addition to glands of this type, there are glands which do *not* have ducts. They allow their fluids to leak through the membranes about the gland cells. The fluids then seep through the thin walls of tiny blood-vessels in the glands and enter the bloodstream. The blood then carries them to all parts of the body.

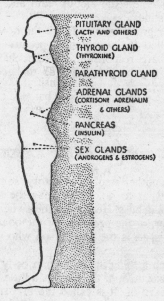

GLANDS WHERE HORMONES ARE MANUFACTURED

PITUITARY GLAND (ACTH AND OTHERS)

THYROID GLAND (THYROXINE)

PARATHYROID GLAND

ADRENAL GLANDS (CORTISONE ADRENALIN & OTHERS)

PANCREAS (INSULIN)

SEX GLANDS (ANDROGENS & ESTROGENS)

Glands such as these are called DUCTLESS GLANDS, or ENDOCRINE GLANDS.

An example of such a ductless gland is the THYROID GLAND, which is located in the throat around the voice-box. Near the thyroid gland are four small knobs of tissue known as the PARATHYROID GLANDS. Above each kidney is a lump of tissue called the ADRENAL GLAND.

These small masses of cells are of key importance to the body. A man can live with one lung removed, or a kidney gone, or his stomach cut out. To remove the tiny parathyroids or the adrenals, however, would mean a fairly quick death.

The pancreas, which we have already given as an example of a gland with a duct, is also a ductless gland. It contains millions of little isolated groups of cells that are called the ISLETS OF LANGERHANS, named after the first man who found and described them. They produce a fluid that contains a hormone and has nothing to do with ordinary pancreatic juice. The fluid produced by the islets of Langerhans enters the bloodstream and does *not* enter the small intestine.

Still another group of glands are the GONADS. They are sometimes called the sex glands, because the hormones they produce are responsible for the changes that take place when a boy or girl grows up to become a man or woman.

THE BODY'S CHIEF EXECUTIVE

We can take another step. In a factory, someone must tell the foremen what to do. There must be some chief executive; some big boss. He may be the owner of the factory, or the manager, or the president of the board of directors. Whatever he is, he is the one person who makes the final decision.

Is there such a thing in the human body? Yes, there is.

The chief executive of the body is a little organ you may never have heard of. It is called the PITUITARY GLAND. It is a tiny thing, weighing only about a sixtieth of an ounce. It is located just below the center of the brain and is attached to it by a thin stalk. In fact, it is located nearly in the middle of the head, so that it may be as protected as possible. After all, it is the key to all the chemistry of the body.

The pituitary gland produces a large number of different hormones. The total number may be as high as twenty-five. At least six of these are known fairly well and have been prepared in pure form.

All the pituitary hormones are proteins. In the form that chemists obtain them, they are very small proteins. For instance, one of the most important is so small that it consists of only eight amino-acids strung together. (Ordinary proteins contain thousands of amino-acids, as you know.)

Actually, protein hormones probably exist in two forms. Inside the cells of the glands, the hormones exist as fairly large molecules. They are too large to leak out of the cell. When the body needs the hormone, however, the protein breaks up into small pieces. The pieces are small enough to pass through the cell membrane and into the bloodstream. It is these small pieces that we are best acquainted with.

Most of the hormones of the pituitary gland act on other glands. (That is what makes the pituitary the body's chief executive. It tells the other glands, which are the body's foremen, what to do.)

For instance, one pituitary hormone is carried by the bloodstream to the adrenal glands above the kidney. Once it reaches the adrenals, it causes them to start manufacturing *their* hormones. The full scientific name of the pituitary hormone which starts the adrenals working is ADRENO-CORTICOTROPHIC HORMONE. This is a long jaw-breaker of a name even for a chemist. So chemists use the initials and call it ACTH. Perhaps you have heard of it under that name. In recent years, it has received much publicity as a "wonder drug" in treating arthritis and such diseases. The 23 amino-acid molecule which has been formed in the laboratory (as I mentioned in Chapter 1) is a form of ACTH.

Other pituitary hormones control the activity of the thyroid gland, the parathyroid glands, the gonads, and so on.

One very interesting pituitary hormone is the GROWTH HORMONE. As you can guess from its name, it controls the growth of the body. Occasionally, a child is born with a pituitary that is incapable of producing enough growth hormone. As a result, he grows very slowly. He may never grow taller than an ordinary three-year-old. Such "midgets," as they are called, can sometimes be seen in circuses or side-shows.

The opposite situation takes place when a child is born with a pituitary that manufactures too much growth hormone. Such a child grows and grows until he is eight or nine feet tall. The circus "giants" are the results of such a condition.

Sometimes, the pituitary suddenly starts producing growth hormone in later life, after a person has settled down to being adult. Most of his bones have hardened by then and can no longer grow. His hands, feet, and lower jaw, however, are still able to grow and they do so. The result is a condition known as ACROMEGALY. People with acromegaly have very large hands and feet, and long chins. Primo Carnera, who was once the heavyweight boxing champion, probably had such a too-active pituitary.

Large animals have large pituitary glands, even larger than you would expect just from their size. Whale pituitaries are enormous and particularly rich in hormones. Whales' pituitaries can actually be used as a good source for ACTH.

How do hormones work? Chemists only wish they knew. Hormones are much less understood than enzymes or vitamins. Chemists don't know for sure how a single one of them works. There are some theories, though, and one of them involves the thin membrane that surrounds every cell.

This cell membrane is composed of protein and of fat-like substances. It acts as a kind of sieve. It allows some chemicals to enter the cell rapidly and some slowly. It keeps some out altogether. Perhaps every different kind of cell has its own type of membrane.

Now suppose a molecule of growth hormone drapes itself over a cell membrane. It changes the nature of the membrane's sieve-like action. More of one chemical can now get through than before; less of another. Suppose that amino-acids can get into the cell more easily when the membrane is covered with growth hormone. The result would be that the cell's enzymes would have a larger amino-acid supply to work on than usual. They would form more protein and the cell would grow and multiply. If growth hormone affected most cells in this way, the whole body would grow.

A molecule of ACTH, on the other hand, would affect only the cells of the adrenal glands. It would drape over their membranes and allow the entrance of more of the substances that the gland uses as building blocks for its own hormones. This would increase the adrenal gland's activities.

THE THYROID GLAND AND FAST LIVING

As we all know, it takes energy to live. The way in which we measure energy is to count CALORIES. Just as we say something is so many feet long or has so many pounds of weight, we say it has so many calories of energy.

For instance, a pint of milk has about 310 calories of energy. A quarter pound of butter has 800 calories. A pound of hamburger has about 1300 calories, and so on.

Now how many calories of energy does a human being use up in a day? That depends on what he does. If he sits around most of the day, taking it easy, and doing nothing more strenuous than typing (as is the case with the author of this book), he can get by with 2500 calories per day. In other words, he can live, if he has to, on two pounds of hamburger a day, provided he gets all the vitamins, minerals, and water that he needs.

If a man has a job that requires a certain amount of activity, he would need 3000 calories each day. If he has a really strenuous job, such as ditch-digging or wood-cutting, he may need 5000, or even up to 8000 calories per day.

But no matter how little work a man does, he always needs *some* energy. An automobile can be stopped completely. It can be parked in the garage and its motor can be turned off. When this is done, it needs no energy and uses no gasoline till it is started again. This is not possible for living organisms. Once life is turned off, it is turned off for good.

Even when you are lying flat on your back, doing nothing, parts of you are still working. Your heart is beating. Your lungs are pumping. Your kidneys and liver are performing all sorts of chemical jobs. It is as though you were an automobile with the engine idling. Not stopped, but idling.

Your body's chemical workings are referred to as its METABOLISM. When you are resting, the rate at which your metabolism is going on is at a minimum. It is the BASAL METABOLIC RATE. More frequently, it is referred to by its initials, and is called the B.M.R.

Perhaps you have had your B.M.R. measured at some time. You lie down on a comfortable cot in a warm, quiet room and rest for half an hour first, so that you can relax completely. It is done before breakfast so that your body isn't working away, digesting its food.

When you are quite relaxed, they put a mask over your face so that you can breathe pure oxygen. They measure the carbon dioxide you produce and from that they can calculate the B.M.R.

The size of the B.M.R. depends on how tall and heavy you are, but the average man has a B.M.R. of about 1700 calories per day. That is the absolute minimum amount of energy he can get along on. A woman, because she is smaller, can get along on 1350 calories, and children need still less.

Of course, that doesn't mean you must eat that many calories every day. You can fast entirely for days on end. But then your body burns up its own fat and tissue to make up for it.

Now it is the thyroid gland that controls the size of the B.M.R. If a man's thyroid gland is too active, he lives too fast. He loses weight. He is very nervous. He can't sit still. His eyes tend to stick out so that he is "pop-eyed." It is as though he were an automobile with somebody's foot lightly pressing the gas pedal while it was parked in neutral. The motor would turn over too rapidly, the cylinder pump too hard and so on.

On the other hand, if the thyroid is underactive, the person tends to gain weight. He is slow-moving and seems lazy.

Now the thyroid hormone (it is called THYROGLOBULIN, and is a large protein) contains substances known as IODO-THYRONINES. These are amino-acids that contain the element IODINE. You are probably familiar with the "iodine" that is used to put on cuts in order to kill germs and prevent infection. That iodine is actually a solution of the element in water and alcohol. Iodine itself is a solid material that forms grayish, metallic-looking flat crystals.

The thyroid is the only part of the body that contains a fair amount of iodine. In fact, in the 1950's, a new way of testing B.M.R. was developed. This involves measuring the tiny amount of iodine-containing amino-acids put out into the blood by the thyroid. The more iodine present, the higher the B.M.R.

Without iodine, the thyroid can't work. When the thyroid doesn't work at all, as in some infants, the child doesn't grow. It is feeble-minded and seems scarcely alive. It is what is called a CRETIN. A cretin can live normally, however, if he is supplied with iodine.

Iodine is quite a rare element. In some parts of the earth, there is so little in the soil and in the crops that people run short of it. In that case, the thyroid grows larger as though it were trying to make up for the iodine shortage by working harder. In that case a person's neck seems swollen, and he is said to have a GOITER.

Goiter hardly ever occurs near the ocean, since seafood contains ten times as much iodine as ordinary land food. Seaweed or "kelp" is particularly rich in iodine.

Nowadays goiter is not much of a danger, at least in the United States. A small bottle of iodine dumped into a city reservoir can supply the whole city with enough iodine to last for quite a while.

In addition, there are many brands of "iodized salt." This is salt with a tiny bit of iodine added to it. If you live well away from the ocean and don't eat seafood very often, it is a good idea to use iodized salt instead of the plain variety if your doctor approves.

People with overactive thyroids can also be treated. One way is to feed them certain drugs which nullify the action of iodothyronines.

INSULIN; THE LIFE SAVER

Perhaps you have heard of the disease DIABETES. Maybe you even know someone with diabetes. It is the most common "disease of metabolism" known. A disease of metabolism is one which is caused not by germs or viruses but by the fact that some chemical reaction in the body isn't proceeding properly.

In the case of diabetes, the trouble is with the islets of Langerhans in the pancreas.

Under normal conditions, the islets of Langerhans produce a hormone called INSULIN. Insulin controls the manner in which sugars are burned to produce energy. Every once in a while, the islets of Langerhans, for some unknown reason, get permanently out of order. They stop producing insulin and sugars are not burned properly. This is diabetes.

In diabetes, the sugars are only partially burned and the chemicals that are produced are somewhat poisonous and can eventually kill a person. Meanwhile unburnt sugar piles up in the blood. This is the best test for diabetes. A doctor can take a few drops of a man's blood and have it analyzed to see how much sugar it contains. If it contains more than a certain amount, the man has diabetes.

There is a simpler test, and one which is almost as good. The blood can only hold the normal amount of sugar. When it is forced to hold more than that, as in diabetes, the extra sugar is always spilling out through the kidneys. It therefore appears in the urine. If a person's urine contains sugar, it is a pretty safe bet that he or she has diabetes.

This test is very quick. Fehling's solution is added to the urine and the mixture is heated over a flame. If it stays blue, everything is fine and normal. If it turns first green and then orange, sugar is present and probably diabetes.

The symptoms of diabetes are loss of weight, unusual hunger and thirst, and more than normal urination. Anyone with these symptoms can find out quickly and easily if he has diabetes. If he has, he can be treated, and live a normal and useful life. H. G. Wells, the writer, lived into the eighties even though he had diabetes.

Since doctors and chemists have discovered that diabetes is the result of the lack of insulin, they know what to do. They obtain crystals of insulin from the pancreas of cattle or swine that are being slaughtered for food. This insulin is combined with other substances and is dissolved in the proper way.

The preparation can then be injected into the tissues. It must be injected and can't be taken by mouth. The reason for this is that insulin is a protein, and if it is taken by mouth, it is digested, and loses its properties.

Only a doctor can decide how much to inject and how often. Once he has instructed the patient, however, the patient can do the injections himself. It is, of course, a wearisome job to have to keep injecting yourself all your life, but it is better to do that than to die. Before insulin was discovered, diabetes patients died.

The diabetes patient must be careful of his diet. You see, when your islets of Langerhans are working, they adjust themselves to conditions in your body. If you've eaten a lot of starch or sugar, they increase the natural supply of insulin. If you've eaten a lot of fat, they decrease it. When you're injecting yourself with insulin, however, you are putting in just one amount, and you must adjust your diet to that level.

Too much insulin can also be dangerous. With too much insulin, the sugar is used up too fast and the sugar-level in the blood drops below normal. When that happens, the diabetes patient usually falls unconscious. To treat such a coma, the patient must be given a sugar injection to use up the extra insulin.

Insulin has been studied more than any other protein. It is quite small, since its molecular weight is only 12,000. The exact order in which the amino-acids are arranged in the insulin molecule has been worked out, just as it has for some of the pituitary hormones. Unfortunately, knowledge of the arrangement has not explained how insulin works.

There are still other protein hormones. For instance, there is the parathyroid hormone which controls the way in which the body uses its calcium. None are, however, as well known as those we have mentioned in this chapter: ACTH, growth hormone, the iodothyronines, and insulin.

SUMMARY

Hormones are compounds that supervise the overall workings of the various enzymes. They are formed in special glands. The most important such gland is the pituitary gland, which forms a series of hormones that supervise all the other hormones. The thyroid gland produces a hormone which contains iodine and which governs the rate at which the body produces energy. This rate can be measured and is called the basal metabolic rate. The pancreas produces insulin, which regulates the manner in which starch and sugar are burned in the body to produce energy. People whose pancreases cannot produce enough insulin suffer from diabetes. They can be helped by being given injections of insulin obtained from the pancreases of cattle. All these hormones are proteins.

10

Other Hormones

ADRENALIN AND EMERGENCIES

WE HAVE MENTIONED the adrenal glands in the previous chapter, but haven't gone into details. As we said, there are two adrenal glands all told, one over each kidney. Each adrenal is actually a double gland, made up of two independent parts.

In the interior of the adrenal gland is a portion called the MEDULLA. Surrounding the medulla is the adrenal CORTEX. The medulla and the cortex produce entirely different hormones, with different structure and different functions.

The medulla produces ADRENALIN. Perhaps you have heard of this substance before. Actually, "adrenalin" is a trade name and the proper chemical name is "epinephrine." However, "adrenalin" is used by almost everybody, even by most chemists, and we'll use it here.

Adrenalin is the simplest hormone known and the first hormone ever to be obtained in pure form. Even the smallest protein hormones have molecular weights of at least 1,000. Compare this with adrenalin, which has a molecular weight of less than 200. Chemically, adrenalin resembles an amino-acid with the acid part knocked off.

In spite of its simplicity, however, adrenalin is one of the most powerful chemicals known. The entire bloodstream of a grown man contains only about two ten-billionths of an ounce. That means that half an ounce would be enough to supply all the people in the world. More than enough.

Two ten-billionths of an ounce is the normal amount. In emergencies, the quantity can go up a thousand times, till blood would contain as much as a ten-millionth of an ounce. You might think that is still an unimportant amount, but when adrenalin goes up that high, you can feel it.

In fact, if you ever get angry or frightened, the first thing that happens in your body is that the adrenal medulla pours a tiny bit of adrenalin into your bloodstream. The adrenalin has a very quick effect on some of the nerves that control important parts of your body. That accounts for all the things that go on inside of you when someone jumps out at you suddenly and yells "Boo."

To understand this, you must realize that there are two kinds of muscles in the body. First, there are muscles like those of the arms, legs or eyelids. They are completely under your control. You can blink your eyes, if you choose, or move your arms and legs whenever and however you want. Muscles such as these are called VOLUNTARY MUSCLES.

Not all parts of your body are like this. The heart, for instance, beats regularly without your having to do anything about it. It beats when you're sleeping or unconscious. You can't stop it; you can't make it go faster or slower.

That's the way it should be. The heart is too important to be left to the moods and fancies of its owner. It speeds up and slows down according to the needs of the body and it does this automatically. The heart is an example of an INVOLUNTARY MUSCLE.

Adrenalin affects those nerves that control involuntary muscles, particularly those that tell the heart what to do. For instance, the presence of even a small excess of adrenalin in the blood causes the nerves of the heart to order it to speed up. The heart pumps faster and harder. It sends out blood, with its sugar and oxygen, circulating faster.

Furthermore, adrenalin also controls the exact organs

to which this extra flow of blood goes. When you are facing an enemy you must fight or a hungry lion from whom you must run, you need all your energy in your muscles. Adrenalin causes the little blood vessels that go to the muscles of your arms and legs to relax. It does this by loosening the involuntary muscles in the walls of the blood vessels. When the vessels relax, more blood can get through them and into the muscles that are going to have to work.

On the other hand, while the emergency is on, the body can suspend such jobs as digestion. The blood vessels leading to the intestines and the kidney contract. Less blood gets through them, so that those organs must cut down on their work. (This is one of the reasons why stomachs sometimes get upset after periods of strong emotion.)

Finally, adrenalin affects the body's chemistry in such a way as to enable it to burn sugar faster.

By doing all this, it is plain that adrenalin is the chemical that takes care of life's sudden emergencies.

THE IMPORTANT CORTEX

The cortex is the really important part of the adrenal glands. This is shown by the fact that the medulla can be cut out of animals without apparent harm. To cut out the cortex would kill the animal. (Of course, laboratory animals live quiet lives in cages. They don't have to face many emergencies and so don't require much adrenalin.)

The word "cortex" comes from a Latin word meaning "bark" because it surrounds the medulla like the bark of a tree. The hormones it produces are called CORTICOIDS.

There are a great number of these. Twenty-eight different corticoids have been obtained in pure form and many more probably remain to be isolated.

All the corticoids have similar chemical structure. Unlike the various hormones we have mentioned so far, they are *not* proteins or amino-acids. The corticoids are steroids.

We mentioned steroids, you will remember, when we were talking about Vitamin D and cholesterol. Perhaps we should repeat a little so that you won't have to look back.

Steroids have molecules in which the carbon and hydrogen atoms are arranged into four connected rings. Attached to these rings are other carbon atoms, and, occasionally, oxygen atoms. All steroids have at least one oxygen atom. Cholesterol, which occurs so frequently in the body, has just one oxygen atom in its molecule. It also has 27 carbons and 46 hydrogens.

The corticoids generally have more oxygens and fewer carbons and hydrogens than that. For instance, the molecule of a typical corticoid would have 5 oxygens, only 21 carbons and only 27 hydrogens.

The different corticoids have somewhat different jobs to do in the body. On the whole, though, they fall into two groups. One group of corticoids acts to control the level of minerals in the blood. The second group supervises the storage of glycogen in the liver. It makes sure that our supply of this starch-like energy-containing substance is adequate (provided, of course, that we are not starving to death). Chemists have fancy names for these two groups of corticoids, but we can call them the MINERAL CORTICOIDS and the STARCH CORTICOIDS.

The two best-known corticoids are DESOXYCORTICOSTERONE and CORTISONE. The second name is all right but the first one is a real mouthful. Don't try to pronounce it. Even chemists have to fall back on the old trick of using initials. Everyone calls it DOC.

Perhaps you have heard of one or the other of these. Both are much in the news lately as "wonder drugs." They represent the two different groups of corticoids. DOC is a mineral corticoid and cortisone is a starch corticoid.

Sometimes a person's adrenals stop working, just as the islets of Langerhans stop working in people with diabetes. In the case of out-of-order adrenals the condition is called ADDISON'S DISEASE. If a combination of DOC and cortisone is used, a patient with Addison's disease can be kept going.

Cortisone has received most publicity in the newspapers and magazines because it is useful in a disease of the joints called ARTHRITIS.

A body's joints are places where two bones meet. When

you bend your arm at the elbow, the bones of the fore-arm and the bone of the upper arm rub against one another.

This takes place easily and without trouble just as a door turns on its hinges. In the case of a door-hinge, metal rubs against metal as the door moves. To keep the motion smooth, the metal must be smooth where the two surfaces meet. In addition, it is usual to put a little oil on the hinge to cut down friction.

The body uses the same system in the case of the joints. Where bone meets bone, the surfaces are very smooth. In addition, the joints contain a liquid known as SYNOVIAL FLUID, which acts to cut down friction.

Now you know what happens to a door when its hinges gradually lose their oil. It becomes harder to open and when it does move, it squeaks. If the hinges get rusty, so that its surfaces become rough, the moving is still more difficult and the squeaking can be horrible indeed.

In one kind of arthritis, the synovial fluid loses its lubricating powers. Joints become painful. For some reason, cortisone seems to improve the conditions of people with this type of arthritis. It has received a lot of attention for this because before this was discovered no cure was known for this very painful disease.

Another hormone which is helpful in arthritis is ACTH. You will remember that this is one of the pituitary hormones. Actually, it is the particular one which orders the adrenal glands to get busy. So you see it amounts to the same thing. ACTH simply gets the adrenals to produce more cortisone of their own, instead of relying on a cortisone supply from outside.

Cortisone is another example of a "wonder drug" that must be used cautiously. If you just read the newspapers, you might get the idea that wonder drugs are as safe as aspirin. You might think that you only have to swallow a few pills and you can cure anything. If you still feel bad, you can swallow a few more. *This is not so.*

We already know that antibiotics can be poisonous if too much is taken. What's more, some people are more sensitive to them than others are.

Cortisone is even trickier. Being a hormone, it is one of the body's executives. One of the things it can do is to affect the body chemistry in such a way that the troubles of arthritis are cut down. But what other effects are these changes in body chemistry bringing about? We can't be sure that they aren't making trouble in another direction. For instance, some doctors have reported that when cortisone is used on patients with tuberculosis, the tuberculosis gets worse. Also mice who are given cortisone seem to catch infantile paralysis more easily.

If you sometimes wonder why druggists won't sell you certain drugs without a doctor's prescription, you should think of things like this.

It is interesting to remember that there are two kinds of hormones. On the one hand we have the protein hormones and on the other, the steroid hormones. These are the only two kinds. Why should that be? Again, chemists can only guess.

Remember that in the previous chapter we mentioned a theory that hormones did their work by sticking to the cell membrane and changing its properties. Well, we said that the membrane was composed of proteins and of fat-like substances. It could be (this is only a theory) that the protein hormones act mainly on the protein part of the membrane. The steroids, which are soluble in fats, would act mainly on the fat-like parts.

Chemists are very interested in just how the body manufactures its steroid hormones. They haven't learned many details yet. It seems quite certain that they are manufactured out of cholesterol, the body's most common steroid. Also, the adrenal cortex is very rich in Vitamin C. When the cortex is busy putting out a particularly large batch of corticoids, it gets low in Vitamin C.

Exactly what the Vitamin C has to do with it, we don't know, but it's a good example of how all the chemistry in the body hangs together.

If we did know exactly how to make corticoids, it would be of great help to medicine. It wouldn't be so expensive to use cortisone for arthritis, for instance. As it is, we have to obtain a large number of adrenal glands from cattle in the stockyards. From these, after much tedious work, we can extract a few crystals of cortisone. No wonder it is expensive.

Lately, a completely new technique is being tried.

Adrenal glands from animals are carefully removed and the blood vessels are kept filled with blood so that the glands are nourished and remain alive. The technique is a difficult one and requires complicated equipment. The main point, however, is that the adrenal's chemical factory is kept working. This process of keeping a part of the body alive by circulating blood through its vessels is called PERFUSION. Naturally, the blood has to contain oxygen and sugar and whatever else the organ needs.

If now we add to the blood which is circulating through the adrenals some steroid, which is fairly easy to obtain, the adrenals change it somewhat into another one. Sometimes the changed steroid that comes out the other end is quite a valuable one. Perhaps some day in the future we can have large factories in which perfused organs of all sorts may be manufacturing many complicated chemicals which can only be obtained now with great trouble and expense.

ESTROGENS AND ANDROGENS

So far, all the chemistry we have discussed in this book has applied to all human beings. Actually, though, there are two kinds of human beings: men and women. And their chemistry is not alike.

At first it is fairly alike. Even their appearances are similar. If you look at a very little baby in a carriage, you usually have to ask the mother if it's a boy or a girl. Even when youngsters start going to school, most of the difference is in the clothes they wear and in the way the hair is cut.

Later on, though, starting in the early teens, they become quite different. For instance, a man is much hairier than a woman. He can grow a beard and mustache. In fact, sometimes he has to shave every day to keep from growing them. A man also has hairier arms and legs and often a hairy chest. (You might think that a woman has more hair on her head but that's just because men cut theirs short.

If men let the hair on their head grow, they could end up with locks and curls just as long as any woman's.)

Another difference is that a man has a low voice, usually, and women have high voices. This is because a man's voice-box is larger. In fact, you can see it sticking out in front of the neck as an "Adam's apple." Men often have quite prominent Adam's apples. Women (and children) hardly ever do.

Still another difference is that a man is usually larger and heavier than a woman, and more muscular too. He has broader shoulders and narrower hips.

Finally, as you all know, women have babies and men don't. For this reason, a woman has to be designed quite differently from a man.

In order to bring about all these differences, special hormones must be involved, one kind for men and one kind for women.

This is exactly what happens. These hormones are formed by the gonads, or reproductive organs. Those that are responsible for the special chemistry of the female are called ESTROGENS. Those for the male are ANDROGENS.

Actually, the gonads of all people produce both types of hormones. Those of women, however, produce more estrogen, and those of men more androgen.

Like the corticoids, estrogens and androgens are steroid hormones. Estrogens and androgens are not very different from each other. You might think that they ought to be very different in order to have one set produce a man and the other a woman, but this is not so.

For instance, the most important estrogen is a steroid called ESTRONE, while the most important androgen is a steroid called TESTOSTERONE. Now the molecules of each of these two compounds is made up of four rings as is the case for all steroids. In estrone, you have an oxygen at the top of the molecule and an oxygen-hydrogen combination at the bottom. In testosterone, quite the other way. The oxygen is at the bottom and the oxygen-hydrogen combination is at the top. That is almost the only difference between the two.

It's as though estrone and testosterone were the reverse of one another. It's as though they were the same compound, but pointed in opposite directions.

Perhaps some of you have seen old-fashioned street-cars with controls at both ends. The conductor used the control at one end till he came to the end of the line. After

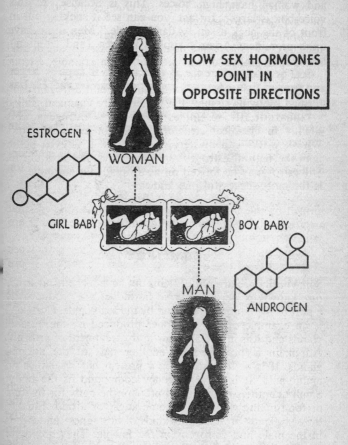

HOW SEX HORMONES
POINT IN
OPPOSITE DIRECTIONS

ESTROGEN

WOMAN

GIRL BABY   BOY BABY

MAN

ANDROGEN

that, he went to the other end of the car and used the other controls on the return trip. Both controls looked the same, but they were pointed in opposite directions. Even though they were the same, using one got you to one end of the city, and using the other got you to the opposite end.

That's the way it is with the estrogens and androgens. Beginning with a pair of babies that can hardly be told apart, one set of hormones changes one baby into a Hollywood actress, while the other set changes the other baby into a champion prizefighter.

Androgens and estrogens, particularly the latter, are used by doctors to fight a number of illnesses. Space doesn't allow us to go into detail here, but, as an example, estrone will help considerably in certain kinds of cancer.

Strangely enough, one of the most useful estrogens isn't formed by living tissue. It is a laboratory chemical, called STILBESTROL. It resembles the natural estrogens only slightly in chemical structure. It is not even a steroid. Why it works is a mystery, but it does. It is even better than the natural estrogens in some ways and it is cheap. Stilbestrol may be looked on as a "synthetic hormone." It is the most successful one known.

SUMMARY

Some hormones are not proteins but have a chemical structure known as steroid. Steroid hormones are produced by a part of the adrenal gland and by the sex glands. Cortisone is an example of the hormones produced by the adrenal glands and it is important now in the treatment of arthritis. Adrenalin is produced by another part of the adrenal glands. It is neither steroid nor protein but resembles an amino-acid. It is the emergency compound of the body. Small quantities in the blood prepare the body for fighting or for running when we become angry or afraid. Finally, the sex glands produce two kinds of hormones: androgens in the male and estrogens in the female. They are quite similar chemically, but affect different parts of the body so that the first causes the development of the bodily structure of a male and the other of a female.

CONCLUSION

In this book, we have dealt with the chemicals that control the workings of living tissue. All are present in very small quantities. All are catalysts. All are important not for what they do themselves, but for what they make other chemicals do. Everything else in the body, left to itself, would be as dead as the rocks about us.

The enzymes, thousands of them, control individual chemical reactions. The vitamins, in some cases, actually form parts of enzymes. In other cases, they may act through their own catalytic powers. Hormones control the overall workings of the body's enzymes.

In this book, you have learned only a very small part of what modern scientists know about these chemicals of life. Yet all that is known today is not enough. The biochemist is forever trying to learn more. Some of the most important problems remain unsolved:

How does a small egg become a human being?

How does the body make proteins?

How do the chemicals of life actually work?

What is cancer and how can it be prevented?

What makes a person grow old and die?

And many others.

If the answer to these great questions is anywhere, it is in more and more study of the chemicals we have described in this book. Perhaps some day, if you are interested enough, you yourself may join the great search.

INDEX

More SIGNET and MENTOR Books

☐ **RED GIANTS AND WHITE DWARFS: The Evolution of Stars, Planets and Life by Robert Jastrow.** A fascinating discussion of the most fundamental questions regarding the origins of the world. (#E7334—$1.75)

☐ **LIFE BEFORE BIRTH by Ashley Montagu.** Advice to the expectant mother on what to do and not to do in order to increase her chances of bearing a normal healthy child. (#Y6590—$1.25)

☐ **THE ORIGIN OF SPECIES by Charles Darwin.** An unabridged edition of this revolutionary classic, with an introduction by Julian Huxley. (#MJ1481—$1.95)

☐ **HEREDITY, RACE AND SOCIETY by L. C. Dunn and Theodosius Dobzhansky.** A revised and enlarged edition of a classic book about the genetic origins of human differences, the use and misuse of the word "race," and the important basic similarity of all human types. (#MT883—75¢)

☐ **THE FOURTH STATE OF MATTER: Plasma Dynamics and Tomorrow's Technology by Ben Bova.** Can the force that gives the sun its energy be harnessed to serve the Earth's needs? This book offers an answer of hope as it explores what may be the most important scientific development of our time—energy derived from plasma, promising to provide our energy needs in the years to come, amongst other important uses. (#MJ1288—$1.95)

THE NEW AMERICAN LIBRARY, INC.,
P.O. Box 999, Bergenfield, New Jersey 07621

Please send me the SIGNET and MENTOR BOOKS I have checked above. I am enclosing $_____(check or money order—no currency or C.O.D.'s). Please include the list price plus 35¢ a copy to cover handling and mailing costs. (Prices and numbers are subject to change without notice.)

Name_____

Address_____

City_____State_____Zip Code_____

Allow at least 4 weeks for delivery

MENTOR and SIGNET SCIENCE LIBRARY Books

☐ **THE SEA AROUND US by Rachel L. Carson.** The outstanding bestseller and National Book Award winner, an enthralling account of the ocean, its geography and its inhabitants. (#MW1317—$1.50)

☐ **THE EDGE OF THE SEA by Rachel Carson.** A guide to the fascinating creatures who inhabit the mysterious world where sea and shore meet—from Maine's rocky coast to the coral reefs beyond the Florida Keys. Illustrated. (#Q4368—95¢)

☐ **THE BIOLOGICAL TIME BOMB by Gordon Rattray Taylor.** THE BIOLOGICAL TIME BOMB explores the dangers that are going to plague our world because of modern day science. (#MW1457—$1.50)

☐ **THE DOUBLE HELIX by James D. Watson.** A "behind-the-scenes" account of the work that led to the discovery of DNA. "It is a thrilling book from beginning to end —delightful, often funny, vividly observant, full of suspense and mounting tension . . . so directly candid about the brilliant and abrasive personalities and institutions involved . . ." Eliot Fremont-Smith—**New York Times.** Illustrated. (#MY1391—$1.25)

MENTOR and SIGNET Ecology Titles

☐ **THE WEB OF LIFE by John H. Storer.** A "first book" of ecology, describing the interlocking life patterns of plants, animals, and man, and emphasizing the necessity of conservation to maintain nature's delicate balance. Illustrated. (#MY1497—$1.25)

☐ **MAN IN THE WEB OF LIFE by John H. Storer.** By the author of **The Web of Life,** a portrait of today's man in today's world—his body and mind, his place in economics, society, politics and civilization as well as the influence he has, for good or ill, on his environment. (#Q3664—95¢)

☐ **THE LIMITS TO GROWTH: A Report for the Club of Rome's Project on the Predicament of Mankind, Donella H. Meadows, Dennis L. Meadows, Jorgen Randers, and William H. Behrens III.** The headline-making report on the imminent global disaster facing humanity—and what we can do about it before time runs out. "One of the most important documents of our age!"—Anthony Lewis, **The New York Times** (#E6617—$1.75)

☐ **TO LIVE ON EARTH: Man and His Environment; A Resources for the Future Study by Sterling Brubaker.** A vitally needed book that shows us where we are on the road to possible extinction. It offers a superbly balanced view of the mixed blessings of our industrial society, the cost of its achievements and what we must be prepared to give up if we want to survive on Earth. (#MW1137—$1.50)

Other SIGNET Books You'll Want to Read